CODE TO JOY

CODE TO JOY

WHY EVERYONE SHOULD LEARN
A LITTLE PROGRAMMING

MICHAEL L. LITTMAN

THE MIT PRESS   CAMBRIDGE, MASSACHUSETTS   LONDON, ENGLAND

The MIT Press would like to thank the anonymous peer reviewers who provided comments on drafts of this book. The generous work of academic experts is essential for establishing the authority and quality of our publications. We acknowledge with gratitude the contributions of these otherwise uncredited readers.

This book was set in ITC Stone and Avenir by New Best-set Typesetters Ltd. Printed and bound in the United States of America.

Library of Congress Cataloging-in-Publication Data

Names: Littman, Michael L., author.
Title: Code to joy : why everyone should learn a little programming / Michael L. Littman.
Description: Cambridge, Massachusetts : The MIT Press, 2023. | Includes index.
Identifiers: LCCN 2022055472 (print) | LCCN 2022055473 (ebook) | ISBN 9780262546393 | ISBN 9780262375979 (pdf) | ISBN 9780262375986 (epub)
Subjects: LCSH: Computer programming—Popular works.
Classification: LCC QA76.6 (ebook) | LCC QA76.6 .L5735 2023 (print) | DDC 005.13 23/eng/20221—dc02
LC record available at https://lccn.loc.gov/2022055472
LC record available at https://lccn.loc.gov/2022055473

10 9 8 7 6 5 4 3 2 1

CONTENTS

1

TELLING COMPUTERS WHAT TO DO
I'VE GOT YOU

If you keep up with the headlines, you know computers are taking our jobs. Spying on us. Controlling what we buy and who we vote for. Even discriminating against us. When they're done beating us at our own pastimes, maybe they'll rise up and kill us. Our relationship with these machines has become, not to put too fine a point on it, dysfunctional.

It was not always so. In the early days of computing, we had healthier boundaries. Back then, by and large, the people interacting with computers were the ones programming them. These people were creating software, building simple games, identifying problems in the world, and figuring out how the computer might be able to help. It was an idyllic time. Unfortunately, that world has gone the way of 8mm home movies.

But we can bring some of its best elements back. It won't be easy, and it will take some time, but we can restore a more natural dynamic. The first step is to remember that these machines are here to empower us.

The second step is bigger: We all need to become programmers.

No, don't freak out! Let me tell you what I mean by the word "programmer."

First, here's what I do *not* mean by it: I do not mean that we all need to learn how to use programming languages the way professional software developers do. That is, we don't all need to learn how to be *coders* . . . although, as a computer science professor, I think it's a great skill to have and it can be very fun.

Here's what I *do* mean: Everyone who uses a computer already tells the computer what they want it to do. We tell computers what we want in ways that range from clicking on buttons to speaking in actual languages, as when we tell Alexa, Google, Siri, or their siblings to "Play us something cheerful" or ask "How long will it take me to drive to Saska-toon?" (Readers in downtown Saskatoon should feel free to substitute a city they would like to drive to.)

But we're highly constrained in what we are empowered to convey, which prevents us from being able to change the computer's response to future events to match our preferences. Being able to do so, in any of several different ways, is what I'm calling *programming*. That's what this book is about. But because the universe of programming interfaces is so huge, I'm not going to go into detail about any particular applications except as examples. And even then I'm not going to attempt to make you proficient in any particular language or interface, such as how to write Excel formulas or how to set the schedule of temperatures on your Nest thermostat.

Instead, we're going to look at what all the particular languages and user interfaces are manifestations *of*: a grammar of concepts that work together to form the fundamental ways we make computers work for us. This topic is particularly timely because methods from artificial intel-ligence, especially in the form of machine learning, are right now trans-forming how computers are programmed by professionals and hobbyists alike. We'll be looking at both the traditional and new ways we commu-nicate our intentions and desires to computers.

And, I promise you, no coding exam at the end.

* * *

There are some people who believe computer programming is a chal-lenging skill that takes a long time to hone. There are others who are just plain wrong. But advances in artificial intelligence and machine learning will help the machines meet us partway. These advances will make com-puter programming more and more like teaching another person. Now, teaching is also hard to do well. But in its most basic form, it's an innate part of the human skill set. We come prewired to teach. Computers come prewired to learn. I think we can make this work.

The goal of this book is to convey how we can tell machines what we want them to do for us so that you, the reader, feel more comfortable tak-ing on those sorts of tasks yourself. Each chapter focuses on a particular element of what can be conveyed, providing examples of how we use these ideas in our daily interactions with other people. The chapters present a few ways you can experiment with these ideas in the computing setting right away, using publicly available systems that might also make you more productive as a welcome side effect. Each chapter also reflects on how bringing these programming components to bear can be expedited using machine learning—the application of data to create useful programs automatically. The various programming components give us the foundation for a roadmap to a future world where programming is more like literacy—something everyone has an *incentive* to learn and the expectation is that everyone *will* learn. I use the term "programming" or "telling the machine what to do" to capture the generic idea of conveying a task that the machine should do on your behalf. It includes coding in the traditional sense of the term as a special case where what you provide is a set of instructions that the computer will follow. But programming includes any way we can convey a desired behavior to a machine to carry out on our behalf. Coding is one particularly technical way to get the job done. But current and imagined future systems take inspiration from the ways we teach each other to do things. Let's take a look at how we do that.

William Arthur Ward, an aphorist writing for *Reader's Digest* in 1965, said:

The mediocre teacher tells. The good teacher explains. The superior teacher demonstrates. The great teacher inspires.

Nice one. Although, as a professional teacher, I have a few issues with this quotation. For one thing, a teacher who doesn't tell you anything is unlikely to be all that inspiring. I'm inclined to think Ward knew that the great teacher draws on *all* of these elements, but, as a great aphorist, he distilled the idea to its essence.

Regardless, I think Ward was really onto something, because the four teaching styles—telling, explaining, demonstrating, and inspiring—work well as a roadmap to the four main mechanisms we have today for telling machines what to do. In what follows, I'll tell (and explain and demonstrate, and maybe even inspire!) you about each of these modes, arguing that we need to be able to draw on all of them symbiotically to make programming more accessible. And making programming something everyone can engage in, creating a kind of universal programming literacy, is our royal road to restoring a beneficial relationship between us and our machines.

WHY WE HAVE COMPUTERS

Computers automate: We give them an information-processing task and they carry it out for us. They are kind of like human brains in that sense. In fact, as recently as the 1970s the word "computer" was a job title that people had.

So, if people were already well equipped to do computing, why do we have all these electronic computers today? Why didn't we stick with the tried and true, and simply have people do all the information-processing tasks? For one thing, people are a lot easier to make. No one has ever gotten drunk in a bar and accidentally built a Dell Inspiron with a graphics coprocessor in the bathroom. And besides, it's pretty hard to explain to a computer-controlled robot how to run a quick errand like picking up frozen yogurt for you from down the street, and make sure there's fruit on top—you know, so it's healthy.

There are several different reasons why we have computers working on our behalf. The kinds of tasks that are better suited to computers are ones at which people are (1) too slow, (2) too easily fatigued, or (3) too expensive.

For example, we have computers on Mars controlling rovers for us. That's because the time it takes for the relevant information to reach a driver on Earth and then for the return signal to actually steer the rover on Mars is ten to forty minutes. Driving a Mars rover from Earth is like playing the world's most tedious game of *Mario Kart*. To make the task less infuriating for the rover operators running the mission, the rocket scientists put computers on board the rover that analyze camera images and help steer the little vehicle toward scientifically interesting rocks (and, I imagine, avoid blue shells).

Occasionally the program gets the rover stuck, and someone in mission control needs to intervene. Would it be better for a Marstronaut to drive the rover directly? A lot better, actually. But we haven't figured out how to keep people alive on the red planet yet, which makes that plan a bit less attractive. Mars really is not the kind of place to raise your kids, as Sir Elton John pointed out. So we have computers on Mars do the work because people are too slow—we can't issue the commands fast enough because we are far away. You could also say we're too easily fatigued since steering the robot for a mile at a speed of 0.002 miles per hour could drive someone crazy. Very, very slowly crazy. And using people would also be too expensive since it's estimated that putting a driver on Mars would cost about a billion dollars, which is even more than an Uber from the airport during surge pricing.

Exploring the surface of Mars is a pretty "out there" application of computation. But the more mundane applications also fall into these three categories. You use computers to send messages because you're just too slow to deliver pictures of adorable miniature llamas to everyone in your group chat. You use computers to find a store that sells your favorite brand of coffee because you don't have the patience to go through a list of all the merchants within a fifty-mile radius. And you use a computer to format a business letter because hiring a graphic designer is not in your price range.

You might have noticed that these three categories are not all *that* well separated. A deficit in any one of the categories can be mitigated with a surplus in one of the others. If you are not easily fatigued, you can make up for a lack of money to do a task by just working longer. If you can work fast enough, it doesn't matter that you are easily fatigued because you'll be done before you are bored. And so on. But we're often lacking in time, energy, *and* money. That's when computers really earn their keep. It's as if they make us all lightning fast, indefatigable, and filthy rich. The last time in history when so many people had access to so much power, it was . . . well, never. In the Middle Ages, there were no apps. You had to be a feudal lord to have a Messenger.

But if computers make us powerful, why is it so painful to use them? Our relationship with computers is, well, complicated. It's not like anyone throws a parade to celebrate when their computer gets an upgrade. But maybe we should. Computer power has doubled around twenty times (making them 2^{20} times more powerful) since the Apple Macintosh was released in 1984. That's a million times more information-processing work computers can do on our behalf today. We just keep getting richer,

Figure 1.1   Miniature llamas are pretty cute, don't you think? *Source:* Photo by Lucas
Andión Montáns. https://www.flickr.com/photos/andion/7249354502.

and faster, and—I don't know, better-looking? The catch is that telling
the computer what we want it to *do* with all that power is hard for us.

WHY COMPUTERS HAVE US

As a result, most people most of the time use their computing devices
for generic things where someone else has already done most of the hard
work of telling it what to do. When we send an email, for example, we
can tell the computer what address to deliver the message to. But we can't
use it to change the email's contents based on when it arrives because
our email app doesn't give us a way to do so, and we don't know how
to do it ourselves. As a result, we are at the mercy of the individuals and
corporations that know how to tell the computers what to do. The relationship we currently have with our machines is not the one *we* want to

have. It's the one *they* want us to have. And they've decided that it's better to have the computers control us than the other way around. You can switch on your robot vacuum cleaner to start cleaning your house, but you can't stop it from sharing with its creator what it learns about you in the process.

Most of these companies did not start off with the goal of insinuating themselves into our personal lives. In 1984, when the Apple Macintosh was announced, the epic Ridley Scott–directed ad said:

On January 24th, Apple Computer will introduce Macintosh. And you'll see why 1984 won't be like *1984*.

Oooh. Gives me chills every time.

You see, back then, computers were just starting to break into the mainstream. They were new and they felt inscrutable and alien. (*Alien* was an earlier Ridley Scott project.) We couldn't understand these devices, so there was no way we could use them effectively. The computers were useless to us without programs, and the available programs weren't that easy to use. At best, we could be given a mechanical recipe to follow to make the computer act on our behalf. It worked, but it cast us as cogs in a giant soulless machine. The *1984* imagery felt all too apt.

Apple set out to bridge the gap between the software and the people who wanted to run the software by making the act of using software *less* like coding. Coding is hard, so if we can hide the complexity, we can make computers less daunting and more people can benefit from the technology. And it worked! Sort of. Apple showed the industry how to make computers more usable by putting software developers in the gap between us and programs. This arrangement enables the developers to make using computers feel like it's not *1984*. But at the same time, it gives the developers irresistible incentives to make it actually *be* like *1984*. Standing between us and the computers gives them power, and power corrupts.

I'm not saying software developers are bad. Some of my best friends are software developers. In fact, I don't think they have much of a choice—the way things are now, they have to stand between us and the machines. That's because people who can't program can't use programs unless those programs hide how they are programmed. Clearly, we need more visibility into what the computers are doing, and that means we need to meet the computers partway. We need to be open to the idea of programming.

If we all start to think a little more like programmers, we can demand that the software developers expose more of the programming of the machines to us. It's extra work for the developers to include "hooks" in their creations that give us more control, so they aren't going to do it unless we (1) show them we're ready and (2) insist on it. Then we'll start to see a shift. It won't happen all at once. But if we can gradually increase our participation in the process of telling machines what to do, we'll begin to restore the proper power dynamic—our computers working for us. The revolution starts today.

EXPLICIT CONTENT FOLLOWS

What does it take to get computers to understand what we want and then do it? We have to keep in mind that computers are the French taxi drivers of the machine world. If you want them to take you someplace popular like the Eiffel Tower or the Louvre, no problem. They do it dozens of times a day, and all you have to do is point to a picture of where you want to go. But if you want to tell them to take you to Coco de Mer, but first swing by your hotel and wait for you to run upstairs to pick up a jacket before heading back along the Seine for a leisurely drive so you don't get there before your date because you don't want to seem overeager, you're going to have to learn some French. The more complicated your request, the better your French has got to be.

That's what this book is about—French. No, no, it's about how we can get the most out of computers by conveying our wishes to them effectively. It's also about the human side of recent advances in artificial intelligence and machine learning that have given us ways for the machines to meet us partway and participate more actively in interpreting our intentions. Programming won't catch on broadly if it looks like coding and requires us to think like machines. It needs to feel more like how we tell each other where to go and what to do.

There are four categories of approaches we can use to tell machines what we want from them. They map nicely onto the four verbs in William

Table 1.1 This 2 × 2 grid summarizes how we tell machines what to do.

	instruction	*incentive*
explicit	**tell**	**explain**
example	**demonstrate**	**inspire**

Arthur Ward's aphorism: tell, explain, demonstrate, inspire. Each is successful in its own way, but each also has significant limitations. We'll see that our best bet for letting all of us share broadly in computation's power is to design future computer systems that integrate the four methods. But even in the short run, this book will give you insights that let you better understand and work with these ever-present electronic helpers.

I've organized these four methods into a two-by-two grid. Right away you can see that you're getting the whole picture because all four cells are filled in. (That's quality book writing right there!)

Don't worry about what's in the cells quite yet. First, let's take a look at the grid's labels. The first distinction, made along the top, is instruction versus incentive. (We'll talk about the vertical labels in the next section.) These words capture two different ways of expressing intent. *Instruction* conveys low-level steps that you want followed. An *incentive* is a way of getting what you want indirectly by coaxing. We use these two approaches with each other all the time. I might ask my daughter Molly to put away the dishes, telling her which dishes go where. That's an instance of instruction: "Do this."

If I'm explicit, this mode is *telling*. It doesn't always go as planned, but we can all agree on what I'm asking her to do. She has to work pretty hard to pretend she doesn't know what I want. Fortunately for her, she's clever and can usually come up with a way of avoiding the chore. "Oh, you meant *today*?"

Incentives are more indirect. They leverage *motivations* for finishing the task and give leeway in *how* the task will get completed. I call this mode *explaining* because we don't give a step-by-step breakdown of what to do but instead explain what we want accomplished. To get the dishes put away, I could change strategy and promise my daughter takeout from her favorite restaurant if she manages to have the clean dishes in the cabinets at the end of each day all month. Now she has a real reason to get the job done. Her love of Thai food can get her to fill in the blanks in a way that works for both of us, although her cleverness can get in the way here, too. "Just so you know, I padlocked the cabinets with the clean dishes inside them. Now that you can't get them dirty, there won't be any dishes to be put away. And please make my pad woon sen mild this time."

People use telling and explaining in the context of computers, too. For example, NASA engineers have two ways to give directions to the Mars rover. In one, the rover is told what to do through explicit instructions:

```
go forward at 0.01 mph for 8 minutes
turn 35 degrees left
wait a day for further instructions
```

In principle, long and detailed enough instructions of this form could get the robot to any target location in one shot. The problem, of course, is that measurements of distances on Mars are not perfectly precise and our knowledge of the obstacles and how they will interfere with the rover's movement is limited. Imagine giving a visitor directions to where you live in terms of distances and angles for them to follow blindfolded. You'll be lucky if they end up in the right city. Or maybe they'll be in the right city but they'll spend the night repeatedly walking into the same temporary police barricade. Not ideal.

Using more sophisticated instructions, the rover operators can convey their intent more reliably. For example, running the wheels at 0.01 mph doesn't mean the rover will actually *move* at that speed. After all, Mars is dusty and wheels can slip on dust, like a toddler in socks sliding on a hardwood floor. If the rover can measure its position and angle independently of the movement of its wheels—visually, say—the rover operators can use a more robust set of instructions:

```
while total distance traveled is less than 0.08 miles:
    go forward at 0.01 mph
while total angle turned is less than 35 degrees:
    turn left at 1 degree per second
wait a day for further instructions
```

The word **while** in these instructions grants the computer a little bit of autonomy. The rover operators don't know how many times the go forward at 0.01 mph instruction will be executed. They are essentially leaving that part up to the rover. But the rover isn't free to choose how far to travel. The instructions are absolutely clear about that.

As a quick aside, even this kind of robust feedback loop can run into trouble. In the 2004 government-sponsored robot race through the desert, the DARPA Grand Challenge, which helped trigger the rush to build self-driving cars, one robotic car followed instructions that went something like this:

```
while target speed is x AND actual speed is less than x:
    increase acceleration
```

That's not an insane thing to do. If the road is sandy and the car isn't making fast enough progress, these instructions rev the engine so it moves more effectively. Which is what it did almost all the time. Unfortunately, at one point the car got stuck in a ditch and its wheels were lifted off the ground. It saw it wasn't moving so it increased its acceleration. Since that didn't help, it increased its acceleration more. Before long, it was moving at 0 mph with the engine running full tilt.

Then it exploded.

That's a taste of what it means to tell the rover what to do with instructions. Lots of detail. Lots of control. Occasional mishaps. But what about incentives? It may seem surprising that we can convey tasks to a rover using incentives. Do you promise it a luxurious oil bath if it does what it's told? It is pretty dusty up there, after all. Or do you go negative and threaten to take away some of its batteries if it doesn't comply? No? So how *do* you motivate a machine?

Obviously, it's not going to be the same as it is with people. Machines don't naturally have urges and desires. We don't cover up our laptop camera because we're about to drink a milkshake and don't want the computer to get envious. Instead, we need to build in artificial motivations and express our tasks in terms of these drives. With this idea in mind, the second way rover operators can provide direction is by specifying waypoints—goals for the robot to reach. It's programmed to satisfy these goals by analyzing the nearby terrain with its camera, considering the safety and efficiency of a range of routes, then choosing the one that is most likely to succeed. It's Martian Waze. The rover "wants" to get to the goal because if it's not at the goal, it is *compelled* to keep working on

Figure 1.2 A rover plans the steps it should take to navigate on Mars. Image created with DALL•E 2 (Open AI).

building and following new routes until the goal is met. The rover operators explain where they want the robot to end up and the robot tries to make them happy.

I want to point out a couple of things here. First, a rover that is working to satisfy an unmet goal might not sound like it has an "incentive." It doesn't really have any choice in the matter. It's not as though there's a bowl of milk at the end of its path, much less that it decides not to procrastinate because it doesn't want to disappoint the rover operators. Still, it's not easy to draw a bright line between human incentives and what you'd find in a more complex version of the target-seeking robot. The robot needs to balance energy usage and goal satisfaction and safety and is making subtle trade-offs. It doesn't get to choose these goals, but it can choose how to satisfy them. Do *we* get to choose what we enjoy and what is most important to us? If you think it's easy to answer this question, I have a few centuries' worth of philosophers who would like to debate the existence of free will with you. I'll bring the popcorn.

Table 1.2 The second row of the 2 × 2 grid focuses on learning to behave from examples

	instruction	*incentive*
explicit	tell	explain
example	demonstrate	inspire

A second thing to mention is that one very interesting difference between incentives and instructions is that a rover following incentives is somewhat unpredictable. The rover operators don't know exactly what route the rover will take to get the job done. And, come morning, if they find the rover flipped over on its back, it may not be immediately clear what the rover was doing when things went wrong. This kind of robot has a richer inner life than the purely instruction-following robot.

Third, in some ways, the challenge of conveying instructions ("left, straight, right, straight, left, left, aaaaand stop!") is much more tedious and error-prone for the rover operators than the challenge of conveying incentives ("be at that rock"). But satisfying incentives makes more demands on the *rover* than following low-level instructions. That's one of the reasons why instruction-based programming—telling—has been so common historically. The computers of yore weren't powerful enough to do anything else.

SETTING AN EXAMPLE

That's the essence of the instruction/incentive dimension of our 2 × 2 grid. What about the vertical explicit/example dimension? The rover scenarios we just went through all involved *explicitly* telling the rover what to do or explaining what to *want* to accomplish. We convey tasks by *example* when the task receiver gets to watch the task transmitter doing the task and it learns the steps, or the goals, by watching.

Instead of another rover example, here's a Rover example. Rover is a dog. I was surprised to learn that Rover isn't a common name for dogs, but the names of my two children both broke the top ten. In retrospect, I think my wife may have wanted a pet and didn't know how to discuss it with me.

How do we teach Rover to sit on command? Since sitting is something dogs do pretty naturally, you could say "sit," and then wait patiently. If you want more immediate results, you'll need to get Rover to make the connection between the verbal command and the target behavior. Explaining what you want in English won't work. Dogs don't speak English, as it turns out. Old English Sheepdogs are the closest, but they only speak *Old* English. Forsooth.

One teaching method that brings better results is to gently push Rover's hindquarters to the ground while saying "sit" and presenting a treat. Nudging his body into the right position means he can associate the taste of the treat, the sound of the command, and the feel of the body position. Instead of describing what we want Rover to do, this experiential *example* of the desired behavior serves as a guide for carrying out the desired behavior. You can think of it as an example *demonstrating* the instructions, or muscle movements, you want Rover to follow. (Note that this kind of demonstration is more direct for the dog than trying to teach Rover to sit by having him watch you cop a squat. There's too much visual-to-motor translation necessary. Robots struggle with these same issues.)

Rover can also learn higher-level behaviors. Once he can fetch a bone in the living room, you can also urge him to fetch an empty cup in the dining room. The visual experiences Rover has and the muscle movements he makes are quite different in the two situations. Rover generalizes because you are *inspiring* him to adopt your goals inferred from examples of successful fetching: "Go to the thing being pointed at and bring it back, independent of what, where, when, and how." He's used examples to extract an incentive, and he can now apply that incentive to guide his behavior in new situations. Surprisingly clever, don't you think? But maybe it's not *that* surprising. After all, these are the creatures that figured out how to wrangle lifetime employment just by being sweet and attentive. That's a trick a lot of people haven't figured out.

Computers can also learn instructions and incentives from examples, using an approach known as *machine learning*. For example, from 2010 to late 2021, Facebook was automatically tagging more than a billion people's faces with their names in uploaded photos. No one gave Facebook instructions to tell it how to tag our faces in photos. It learned to associate images and names from examples of tag-image pairings users made in the past. It learned from demonstration. Then, when a new photo was posted, Facebook proactively applied what it learned, putting names to faces. Amazing! And amazingly creepy. (Facebook discontinued the practice in November 2021 in response to widespread concerns about misuses of facial recognition technology.)

To sum up, the four principal ways we tell machines (and people, and animals, and . . .) what we want are:

- Telling (giving explicit instructions): "Take these steps."
- Explaining (providing explicit incentives): "Try to make this happen."
- Demonstrating (giving example instructions): "Watch what steps I'm taking, and then take them."
- Inspiring (providing example incentives): "Watch what I'm trying to make happen, and then try to make it happen."

Each approach has strengths and limitations, which we'll talk about throughout this book. Using these techniques *in combination* is often the best way to get across what you want. You might notice, for example, that I use a combination of examples and explicit verbal descriptions to try to convey these concepts. Oops, I just did it again.

I use this structure not just because I'm practicing what I'm preaching. It's because that's how we communicate. It's how we convey information, it's how we teach each other, and it's how we should convey tasks to machines. A problem with machines is that we too often design them to take instruction in only one mode at a time, and that can be awkward and cognitively taxing. Mixing the modes together—telling and explaining, demonstrating and inspiring—is how we best make ourselves understood.

In the following chapters, we'll talk about expressing tasks using different elements of programming (telling) but also how to get across analogous concepts using machine learning (explaining, demonstrating, and inspiring). That will help us better recognize how these ideas can be blended together to make computers more helpful and give us some confidence that it's something we would be willing to do, if the software developers give us the chance.

It's a major shift to go from a world where only professionals program to one where nearly everyone programs. Making this kind of power broadly available requires a significant amount of infrastructure, and sometimes I worry that we might not get the support we need. One encouraging story that suggests that this kind of powerful technology can become widespread comes from fifteenth-century Korea.

GREAT SCRIPTWRITERS

In 1418 the king's son Sejong turned twenty-one years old and became the new ruler of Korea. He was a creative leader and a lover of science. He helped build a more egalitarian society, including by promoting non-nobles to positions of power and influence. Although he worked closely with the Ming dynasty in China, he initiated updates to the Korean calendar system and Korean medicine, shaping a more independent Korean identity.

At the time, Korean was written in Hanja, a system consisting of tens of thousands of Chinese characters. Hanja was so difficult to learn that only privileged aristocrats, usually male, could invest resources in developing the skill to read and write fluently.

Sejong combined his interests in sharing knowledge, science, literacy, and Korean uniqueness to invent a new writing system. Instead of requiring users to memorize somewhat arbitrary mappings from characters to words, his system connected directly to how words were pronounced—something that everyone who could speak the language would be able to understand. Further, the symbols themselves were chosen to reflect how their corresponding sounds were made, making it easier to pick up and remember the relationship between symbol and sound.

All told, Sejong created about a dozen consonant symbols and a similar number of vowel symbols, and devised a way for them to click together like little Lego pieces to make syllables. Now, your typical invented writing system is unlikely to catch on, what with being created by a geeky middle schooler or obscure academic philologist. But Sejong's system had a notable advantage. Sejong was the king. So he wrote up his ideas and proudly presented the system to his courtly advisers.

Initially, the scholars and government officials resisted because they didn't see anything wrong with the way they had been doing things. Besides, what do commoners need to read and write for? They said the

Figure 1.3 The author visited King Sejong's office in Seoul where Hangul was created. Author's photograph.

new writing system was a waste of time, harmful to learning, useless to the government, and nothing good would ever come of it.

Nevertheless, good things *did* come from it. Ordinary Koreans learned Hangul, which translates to "great script," enabling them to record the tricks of their trades, passing along the wisdom they acquired so that others could benefit. They could write each other letters. They could file written claims with the government. Over time, poetry, trashy novels, and personal diaries used the new alphabet. No surprise to us, broad literacy had broad impact.

There are many people pushing for programming literacy today, and some of the criticisms sound an awful lot like Sejong's naysayers. They

Figure 1.4 When "Hangul" is written in Hangul, each symbol takes the place of a component sound.

argue that we already have plenty of bad coders and that making more of them, especially ones who use "dumbed down" computer languages that are easier for people to pick up, simply makes things worse. One of my colleagues out and out told me that my approach was "irresponsible."

To be fair, I think we're all talking past one another. They are talking about the impact of broad programming literacy on programming professionals. I'm talking about casual, everyday programming that empowers people to use computers to express their individuality. It can be used as a creative outlet or as a personal problem-solving tool, and I think it's a win.

Imagine if product designers could count on everyone being able to do basic programming in whatever a future analog of Hangul might be. Would we have more powerful products? Would doorbells and washing machines and toaster ovens and car dashboards be different? I don't know. It seems to me that in the old days, when the main consumers of software were programmers, the programmability of end-user software was well supported.

Throughout the book, we'll look at examples of systems that support some degree of programmability—universal remotes, questionnaires, home automation, spreadsheets, video-game builders, Google apps scripts.

Each of these examples has its own idiosyncratic syntax for specifying behavior. As a result, the only transfer that people get from one to another is a generic ability to think computationally. That's good, but it's not great. In one honors seminar I asked students to document how to change the time on their clocks. Thirteen students came back with thirteen different procedures. Returning to the reading example, it seems analogous to the awkward situation we'd have if there were a different alphabet for each application of writing (passports versus clothing-care instructions versus movie subtitles). People would hate that, probably as much as people hate changing the time on their clocks.

Pulling these ideas together, here's my vision for a world where people can tell machines what to do:

1. A common syntax would exist across applications and products and be optimized for simplicity and applicability (much like Hangul in the writing case).
2. The need for true experts to create programs that are consumed on a large scale would remain (just as there is still a role for authors, poets, and journalists in the writing case).
3. Mundane, and currently unforeseen, personal applications would come into being that depended on programming knowledge (such as making to-do lists and writing love letters in the writing case).
4. A widespread desire to learn programming skills to participate in these applications would emerge (analogously, the importance of learning to read and write is unquestioned, even in our highly polarized world).

I want to contrast this last item with a standard motivation advocates give for teaching programming: "It helps you to learn to think computationally." That strikes me as a bit of a circular argument: Learn to program so you can think computationally, which is key to being able to program! I don't think it works as a motivation. There need to be concrete, real-world benefits, which programming could have in spades. We just need to highlight them better.

The vision hasn't been realized yet, of course. My goal in this book is to teach you how to tell computers what to do today. Instead of focusing on any specific language, I'm going to show you the underlying grammar of programming languages.

Why? Because:

1. Maybe, even pre-vision, you'll want to learn one of the specific languages, say, Microsoft Word's scripting language. What I'll teach you will make it much easier for you to do that.

2. There are signs that we're getting closer to making the vision a reality. For example, user-programmability is already important in some computer games, and powerful application programming interfaces (aka APIs) are increasingly available that let you extend widely used systems such as the messaging service Slack. Helping to get you ready to take advantage of these inroads is important for encouraging progress toward the vision.

3. Finally, even if the vision is never fully realized, you'll leave this book with a far better understanding of how computers work, which is the most important first step in regaining control over them.

So let's get to it!

2

THE WHAT OF PROGRAMMING
ALLOW ME TO INTRODUCE MYSELF

When I say we need to learn how to tell machines what to do, I have something particular in mind.

Pressing the off button tells the computer to power down, tapping the volume button on the remote tells the TV to become louder, and selecting the slanted "I" from the menu bar tells the text editor to *switch to italics*. I think of these as examples of using the machine, not telling it what to do. You are making the machine an extension of yourself and simply making it *do* something.

That by itself is pretty remarkable. We human beings can expand our sense of self to include machines. We don't just *use* tools, we *absorb* them. Crane operators aren't thinking about how their hands are manipulating the joystick but of how the shovel is manipulating the gravel. Similarly, when our conscious attention is involved in carrying out a task through the computer, it's we, a kind of computer-human hybrid, that is doing the work. I call this mode of interaction "control."

The focus of this book is on a different activity. It's about conveying to the machine something it should do in the future. This mode of interaction is generally known as programming. Usually programming means coding—writing programs made up of lines like this:

```
if newdist < bestdist: (bestdist, bestid) = (newdist, id)
```

However, I'm using the term more liberally to mean "conveying to the computer how you want it to behave." It includes coding as a special case, but can also include conveying a behavior by means of example. Basically, anything in chapter 1's 2 × 2 grid is fair game: telling, explaining, demonstrating, and inspiring are all ways of programming.

You might want to object that I'm using the word in a way that stretches its meaning too far. I counter with etymology. The word "Program" has Greek origins. The "pro" means *before*, as in words like "prognosis" (before + know) and "pro bono" (before + being part of Sonny and Cher). The "gram" means "writing," as in words like "anagram" (rearranged + writing) and "Instagram" (sharing pictures is easier than + writing). A *program* is what you write beforehand that lays out your plan of action. We still see that sense of the word today when we look in the program for a theater performance to see what the scenes will be. I'm using "program" in precisely this way—as the thing that's worked out ahead of time.

Consider coordinating a surprise party with a group of your friends. You say "1, 2, 3" and everyone yells "Surprise!" The "1, 2, 3" is telling everyone to say "surprise," but the only reason they knew what that meant was because of the plan you put in place in advance. Perhaps you *told* them what to do: "When the time is right, I'll say '1, 2, 3' and you all yell 'Surprise!' together." Or maybe you gave a *demonstration* for them to imitate: "Watch me: 1, 2, 3, Surprise!" You could have also *explained* what the group is trying to achieve: "He doesn't know that you are all here and I want to make it as exciting as possible. When he is about to arrive, I'll give you a countdown and we'll all shout out 'Surprise!' together." You could even have *inspired* them with a deeper appreciation of what they are trying to accomplish by playing YouTube compilations of surprise party entrances. (Some of them are pretty hilarious, actually.) The focus in this book is on the act of preparing for—programming—the big event, not the signal to act—control.

When I'm feeling a little grandiose, I refer to programming as "delegation of the will." To the extent that we associate the word "will" with our ability to form intentions and enact them in the world ("free will," "fire at will," "will they or won't they?"), controlling a machine is enacting our will. In contrast, programming is taking that will, that impetus to act, and installing it in the machine itself. You no longer need to be an active participant because the machine has become a vehicle for your will.

You already program machines in this broader sense of the word. When you select a number to dial from your contact list, you are exerting control, but adding a number to your contact list for later use is a simple act of programming. Calling an Uber is an example of control, but putting your credit card information into the app's settings so you can pay with the touch of a button is programming. Calling an elevator using the "up" button: control. Selecting a floor once you get into the elevator: programming. (If elevators were more like buses, where you have to wait for your destination and pull a cord to request a stop, then it would be control.) Making coffee: control. Setting the timer on the coffee machine so it makes coffee to help you wake up: biological necessity.

The dichotomy is between doing something now for an immediate result and setting up something to happen later. The first is control. The second is programming. But there are boundary cases. For example, turning on the light so you can see is control. But leaving the light on so late-arriving Motel 6 guests can see, well, that's programming.

Form filling in your web browser is one last interesting case worth mentioning. Typing your name into a slot is control. But setting up the browser so it can autofill the information in the future is programming. The interesting part is that there are several ways to do it. You can just fill in a form and let the browser learn from the example (demonstrate) or you can go into the preferences interface and fill in the information there (tell). There are interfaces to auction websites where you leave it up to the computer to fill in your bid after you've told it how high you are willing to go (explain), but if those values were inferred by the machine watching how you spend money on other sites, that would be risky (inspire).

I admit that the boundaries here aren't perfectly crisp, but I find the control/programming dichotomy pretty useful. At the very least, you could think of it as a spectrum. At the control end, you are watching and making decisions continually throughout the process. At the programming end, you make your decisions up front, then set it and forget it. Of these two parts, it's the "set it" that we're focused on here. The second part . . . actually, I don't even remember what that was.

Let's take a look at an extended example of an everyday task where better programmability could have a positive influence on our relationship with computers.

TMI

Academics share ideas by publishing them. In some fields, such as history, that means writing books. In others, such as economics, it means writing journal articles. In computer science, our primary publication outlets are annual conferences, and the fight to get a paper accepted for presentation at a leading conference can be fierce. At a typical top conference, for every five or six papers submitted for review, only one gets published in the conference proceedings.

But look what's happening in the field of artificial intelligence. The Neural Information Processing Systems conference, or NeurIPS, is one of the most respected conferences in machine learning. Submissions to the conference doubled from 2014 to 2017. They then doubled again from 2017 to 2019. The growth of submissions to the Association for the Advancement of Artificial Intelligence conference is even more spectacular: It had a little more than one thousand submissions for the first time in 2014, but for the 2020 conference, just shy of ten thousand papers were submitted.

I was one of the reviewers for the 2020 conference. That meant I had to flag a set of those papers that I'd be well suited to review. To do that, I needed to identify fifty of the most relevant papers out of the ten thousand available. It's quite feasible to skim through one hundred or so titles and mark the ones on topics I'm well informed about. But skimming ten thousand? That's the opposite of feasible. If only there were some way to get a computer to help identify the papers that are likely to be a good match for me!

Identifying papers to review is one example of a more generic problem: Can we get a computer to help sift through many, many items to identify a handful that are worthy of a specific person's further attention? Another name for this problem is *recommendation*. I think it's a particularly acute point of conflict in our unhealthy relationship with computers. The amount of information we can access—news briefs, video clips, collections of gluten-free blueberry muffin recipes—far exceeds the amount we can go through. Without a computer intermediary, the situation would be hopeless. Tech companies offer to help us with this problem, but what we get is *their* decisions about what is good for *us*. For example, some recommendation systems are told to recommend items that are likely to generate "clicks," but that can lead to them prioritizing content that elicits negative emotional reactions, even if that's not what we want. We need to be able to program our own recommendations. So let's take a look at how computer-mediated recommendations work.

In the 1960s, library card catalogs were being digitized, and so librarians were the first to look at this problem in a serious way. By the 1990s the number of items people had to sift through was skyrocketing and the conversation shifted from "retrieval" of relevant items to "filtering" out irrelevant items, but both are forms of recommendation. Today, recommending items to people is big business. That's the main thing that Google does, and it is the fourth largest company in the world by market capitalization.

For a computer to make recommendations, there are a few prerequisites:

1. The computer needs to have information about all the possible items.
2. The computer needs information about what sort of items *you* are hoping to get.
3. The computer needs a method for assessing how well each item satisfies your wishes.

How we tell the computer what we want (#2)—depends on how the computer assesses how well each item satisfies what we want (#3). One common approach is known as "collaborative filtering," which, generically, means people working together to help each other identify good items. To understand how our relationship with computers has shifted over the years, it's eye-opening to look at how the specific meaning of "collaborative filtering" has changed.

Google Scholar is an example of a recommendation system that suggests relevant academic papers for your needs. I asked it for papers on collaborative filtering and the two papers it judged most appropriate are papers on software systems that their designers described as performing collaborative filtering: Tapestry from 1992 and GroupLens from 1997. Just five years separate these two influential papers on collaborative filtering, but the meaning of the phrase shifted dramatically during that time.

Tapestry was designed as a shared repository of timely messages, including group emails, news articles, and posts on the pre-web NetNews—a kind of OG Reddit. (The main difference between NetNews and Reddit is that some NetNews newsgroups turned into what Georgia Tech's social computing expert Amy Bruckman called "a flame fest filled with misogynist hate." Actually, that's not too different from Reddit after all.) Recommendations were made to users based on the queries they wrote in a system-specific language. For example, if a Tapestry user created the query:

```
(sender is "David" OR date is on or before August 31, 2021)
    AND subject contains "contract"
```

the Tapestry system would keep an eye out for new messages that were either sent by David or were sent before September, 2021, but only if the subject line of the message contained the word "contract." If that seems a tiny bit cryptic to you, I would have to agree. These OR and AND operators and the parentheses aren't familiar to most people and take some time to decode. In the early days of electronic card catalogs, such logic-based queries were common, but over time, it became clear that many people found inputting them off-putting.

Tapestry let users give names to their queries, for two reasons. First, it enabled users to re-run the queries just by typing in their names—something we'll look at more closely in chapter 7. Second, Tapestry in this way made an interesting type of collaboration possible because people could make queries that referred by name to other people's queries. So, if Aylin wrote a query that you really liked and gave it a name, such as _fairness_, you could incorporate it into your own queries:

```
message matches Aylin's fairness query
    AND message body includes "reinforcement"
```

That query would flag any message picked up by Aylin's fairness query, but only if it had the word "reinforcement" in it. The fact that you and Aylin could work together in this way to build your recommendations is why the designers called it a collaborative filtering system. So far, so good? The 1992 system made recommendations based on logical expressions, a kind of program, written by individuals and shared with other users.

The later system, GroupLens, is based on the same high-level notion of collaborative filtering—the work other people do to identify useful information is leveraged to help _you_ find good stuff. But GroupLens formalizes the problem very differently from the way Tapestry did. Whereas Tapestry was about coding your own filter, GroupLens learns the filter for you. In terms of the 2 × 2 grid in chapter 1, we've moved from "telling" to "demonstrating."

GroupLens, based at the University of Minnesota, recommends items to users by leveraging its database of ratings—how much each person likes seeing each item GroupLens shows them. It makes new recommendations in two steps:

1. For the user for whom a recommendation is needed, find other users whose ratings of items match closely that user's ratings. (First, search for like-minded people.)
2. See what those similar users thought of each potential item and average their ratings accordingly. (Use those people to identify desirable items.)

This powerful idea, introduced in GroupLens, has been very influential. Companies such as Netflix and Amazon and TikTok use variations on this idea to make recommendations to millions of people each day.

Returning to the 2 × 2 grid from chapter 1, recommendation systems such as Tapestry began by requiring people to *tell* computers what they wanted using explicit logical queries. That gave people a lot of control, but it was also burdensome for them to express every detail of what they wanted. Designers responded by moving dramatically in the other direction in systems such as GroupLens. They began tracking people's every click. The resulting data serve to *demonstrate* the choices people make. It can also *inspire* the machine, in the sense that the computer can adopt people's inferred preferences to provide what they want without their having to spell it out. That's less work for the individual, but it also means users have less say about what recommendations they get.

The success of automated approaches to collaborative filtering helped lead to their widespread use in recommendation systems in the tech industry. As I see it, Google alone deploys recommendation systems for music, movies, restaurants, toys, videos, news, scientific articles, books, images, and, oh yeah, web pages. In many cases, notably in recommending YouTube videos, a recommendation system can leverage "revealed preferences." That is, instead of asking users how much they liked a video, the system makes a guess based on how often users that were given the chance to click on it did so and how long they watched it before switching it off. Using revealed preferences is very clever and shockingly accurate—and a far cry from the challenge of writing detailed logical queries.

A major concern with revealed preferences, though, is whom they are revealed to. Big tech companies use our tracked clicks to create intricate models that can predict our choices with uncanny precision. But our choices and what we want are not always the same thing. Yes, I clicked on an ad for the movie *God's Not Dead*, but it was out of a morbid sense of curiosity about whether it would portray the "atheist professor" character with understanding or at least compassion. (It would not.) I did not want to be barraged with ads for the movie or its three sequels, which, if I recall correctly, were *God's Not Dead: He's Just Working from Home out of an Abundance of Caution*, *God's Not Dead: But He Told Me to Keep an Eye on His Stuff until He Gets Back*, and *God's Not Dead: But You're Going to Be if You Keep Pestering Me about It*. The profile the streaming service built that says I'd likely click on *God's Not Dead* again given the chance is probably correct. But truly, I don't want the service to send me more of those ads. I don't want my moments of weakness to define me, digitally or otherwise.

Since the mathematical prediction models the companies build about us aren't revealed to us, we can only guess whether they are aligned with our true desires, and our hands are tied if the alignment is off. The recommendations are *about* us but not *for* us. They are optimized for our subconscious impulsive selves that skim and click and react, and not for our more reflective and rational selves that have long-term goals and can make subtle trade-offs. That's not good.

With the explosion of information available online, getting recommendations that work for us individually and socially is perhaps the most important task machines do for us. As such, it is one of the most important places where we need to have control. Selecting which items a person sees colors that person's perception of what is possible. It colors the person's perception of reality. This arrangement gives machines, and the companies that design them, outsized power and responsibility.

I would like to see future recommendation systems incorporate the fourth quadrant of the 2 × 2 grid by providing options for users to *explain* their preferences so the systems can tune their behavior accordingly.

For example, some existing sites let people write reviews. What if the recommender systems read the reviews to understand the justifications for a user's rating—and then recommended items based directly on justifications instead of just guessing what users want to see based on the scores? Justification-based recommendation could fit into the space between writing logical expressions (telling), as in Tapestry, and purely observation-based recommendations (demonstrating/inspiring), as in GroupLens. Ideally, it would better capture our true preferences and better align what the system gives us with what we actually want. That would make the internet healthier for all of us. But it requires that we have a richer toolbox of ways to tell machines what we want from them— better ways to program. Having machines simply watch our behavior and infer our goals makes sense as part of a portfolio of methods and not the sole way that systems interact with us.

LEARNING TO PROGRAM, TEACHING TO PROGRAM

I like this quotation from Ben Shneiderman, Distinguished University Professor in the University of Maryland Department of Computer Science, which has been used to argue that we need to teach coding—and teach it well—not just to people who will program for a living but to a much larger swath of the population who would benefit from the power that computers would provide them:

For every professional programmer there are probably ten occasional programmers who write programs for scientific research, engineering development, marketing research, business applications, etc. And finally there are a rapidly growing number of programmer hobbyists working on small business, personal and home computing applications.

It's not just about programming as a job but programming as an enabler of other activities. I'd like to say that his insight will help bring about broad coding literacy any day now, but that's probably not realistic. After all, this passage comes from a book published in 1980. If we knew more than *forty* years ago that coding is an important skill, why is it still a struggle to get people to do it? I think there are two big reasons: (1) coding is hard to learn, and (2) we haven't yet tapped into people's natural ability to teach.

That developing expertise in coding requires a significant invest-
ment of time is not that shocking. Coding takes place in a computer
language. We know that learning languages is challenging because it
takes kids a while to pick up their first language, and acquiring a second
language in school is not always that successful. For example, I've had
way more semesters of Spanish than calculus. I consider myself quite
proficient in calculus, but my Spanish is still *no mucho excelente*. Worse,
languages are for talking to other people, so when we're learning a new
language, it's more about how to say something than what to say. We
have a pretty good sense of what's going on in other people's heads.
Not so with computers. Computers aren't just from another culture or
even another species. They are an alien life form. I mean, what are they
thinking?

Back when Shneiderman was starting to articulate the idea of "software psychology"—understanding the human factor in computer
usage—Elliot Soloway and colleagues at Yale were starting some really
interesting research trying to document the rocky road people take on
the way to learning to code. One result from this work is still being talked
about today. Soloway asked undergraduates taking their first or second
programming course to write code to solve the following task:

Write a program that repeatedly asks the user for numbers until it reads the
number 99999. After reading 99999, it should print the correct average. That is,
it should not count the final 99999.

Variations on this task have come to be known as the *rainfall problem*, because into every educational experience a little rain must fall.
Not really—it's because the problem comes with a backstory where the
program accepts measurements of the amount of rain falling over a set
of days and then reports the average. Soloway expected people to break
down the problem like this:

1. Keep a running total of the numbers encountered.
2. Keep a counter of the number of numbers encountered.
3. Ask the user for a new value.
4. If that value is 99999, output the average and stop.
5. Otherwise, add the new value to the running total and add one to the
   counter.

In code, that could be expressed as:

```
set running total to 0
set counter to 0
input a new value from the user
if new value is not 99999:
    add new value to running total
    add 1 to counter
    loop back to input
otherwise:
    output running total divided by counter (the average)
```

This code is a relatively simple combination of the fundamental programming components that a beginning programmer would be familiar with. The overall flow from one statement to the next is a sequence of *commands*. The running total, counter, and new value are stored in *variables*. There's a *loop* that takes control back to the input to repeat some operations. A *conditional* check determines whether or not the value the person entered matches the "end of list" value 99999 and changes the behavior of the program accordingly. Finally, the entire task can be thought of as a newly defined *function* for carrying out an operation that might be useful in other contexts. We're going to look closely at each of these five elements in the coming chapters. For now, though, the interesting upshot is this. Students taking a formal class in programming struggled mightily with this exercise. Only about 40 percent of the students could do it, with the others getting tangled up using the variables inconsistently, leaving the conditional outside the loop, or forgetting to get new inputs. These highly motivated, bright, deeply invested students bombed.

But don't despair. I don't think it's that they didn't know what to do. I'm pretty confident they could have expressed the task sufficiently well to another person. But breaking the task into a stand-alone description in terms of unfamiliar little steps, each of which needed to be selected carefully and written precisely in a new language, was tricky.

I'd argue that, despite the challenges people face learning to program, teaching others to carry out complex tasks is actually quite a natural activity. One evolutionary theory claims that our ability to convey complex tasks to each other co-evolved with our ability to create complex tools. The basic idea is that fancy tools were essential to survival for early humans, and these tools were sufficiently unlikely to be discovered from one generation to the next that an explicit capacity for passing the skills to other people was beneficial. That is, our ancestors who were born with an edge in explaining complicated tasks were better at helping their off-spring and other community members thrive. Survival of the teachers! Over time, our species became better and better adapted to teaching, ben-efiting all of us. It kind of makes me wish our society held schoolteachers in higher regard. I know we're not all paragons of explanatory brilliance and we often lose patience, but I'd argue that we all have a gift for teach-ing. We take to it naturally, especially when we sense a need to know in our prospective pupils, and we're relatively effective. People may differ in their interest and proficiency in teaching, but we're all a heck of a lot better at it than goldfish, wolves, or chimpanzees.

Kids as young as three and a half have been shown to engage in effec-tive peer teaching in laboratory experiments. They can recognize that they know something that their partner doesn't know and fill in the gaps in their partner's knowledge so that the two of them can work together to accomplish a predefined task. By that age, they have seen lots of exam-ples of being taught, so to some degree, they are imitating what they've experienced. But our innate ability to see others through the lens of "I've been in that situation and I know what I needed to hear to become better myself" is what makes it possible for us to help other people up the ladder of understanding. It's pretty remarkable, even if rarely remarked upon.

I even think the phenomenon of "mansplaining" provides indirect evidence of our natural abilities to teach. After all, the reason we find it so infuriating when a man talks to a woman as if the woman doesn't already understand something is because *usually* other people are pretty good at assessing what others know and what they need to know. So when Jacob assumes a woman has to be told what a virus is even though the woman is an immunologist, it stands out as a gender-based exception to our spe-cies' general strength as teachers. Jacob is missing a key competence, and it's fair to blame him for it.

We're generally natural-born teachers, and that suggests we should be in a perfect position to get across simple tasks to machines. The fact that computers can't put it together and produce the behaviors we want them to carry out based on how we want to teach them . . . well, computers are missing a key competence, and it's fair to blame them for it.

It's not all on us to learn how to tell computers what to do in the form they demand. Computers need to bring more of their power to bear to tap into our natural abilities to teach them. Computer scientists should be helping to bridge that gap by designing more naturally teachable machines. Progress is being made, but not enough and not fast enough.

You and I can't do much about that directly. But we can hasten the day by learning how to tell computers what to do until they've gotten better at listening. So here we go.

3

SEQUENCING COMMANDS
I'LL TAKE YOUR ORDER NOW

When I was in elementary school, my parents sent me to an overnight camp for Jewish kids who liked sports. So, natural athlete that I am not, one of my favorite activities was square dancing. It was awkward, but I liked the structure of it. The caller would teach us the basic moves: circle left, bow to your partner, bow to your corner, allemande left, promenade, do-si-do, and so on. Then, as a song played, the caller would tell us which step we should take. All I had to do was wait for the command, do the step I was told, wait for the next command, repeat. It made dancing as easy as Simon Says. Yee haw!

Line dances at weddings have a similar vibe. I struggle with them, but my wife and kids are great and always join in at festive gatherings. I do better with some particular songs where the lyrics let you know what you need to do. The song "YMCA" tells you what letters to make when. The Isley Brothers' "Shout" not only tells you what to do, it provides fine-grained instructions concerning the volume at which to do it. All of the lyrics of "Cha Cha Slide" tell the dancers what steps to make and the beats to do them on. There are instructions about what direction to step, when to hop, how many times to stomp, and a general exhortation to get funky with it.

I can muddle my way through these songs at a bare minimum level of getting funky with it. But other songs, like "Macarena," have dances that you just have to know. The sequence of commands that make up the Macarena dance are:

Extend right arm straight out in front, palm down
Extend left arm straight out in front, palm down
Flip over right hand, palm up

Flip over left hand, palm up
Move right hand to left shoulder
Move left hand to right shoulder
Move right hand to the right, back side of your head
Move left hand to the left, back side of your head
Place right hand on left hip
Place left hand on right hip
Place right hand on right hip
Place left hand on left hip
Wiggle hips
Jump up, making a quarter turn to the right

If you don't speak Spanish, you might think that's what the lyrics are telling you to do. They are not. If you translate them, all you will learn is that we're encouraging wedding guests to celebrate some pretty racy stuff. (Macarena's boyfriend joined the army so she hooked up with two of his friends? Yikes.) Since the songs don't tell people what to do or when to do it, dancers have to come with the sequence of moves already in their heads. If you can dance the Macarena, it's because you learned the instructions and stored them away for later use. Remembering the steps and carrying them out in order is a challenge (for me) that is unnecessary in square-dance-style songs like the modern "Cha Cha Slide" or the classic "Hokey Pokey."

The steps for the macarena are a good example of the first and most basic mechanism we have for conveying tasks to machines: command sequences. In the case of line dances, we are playing the role of the machine carrying out the steps, and the choreography is the command sequence we are executing. In the terminology of chapter 2, it's programming, not control. The most important element is the instructions themselves, which I'm calling the *commands*. The machine and the dancer need to know the set of possible instructions and what they mean. In:

    take 4 grapevine steps to the right

if the dancer doesn't know what a grapevine step is or can't carry it out, the whole affair is a nonstarter. The same is true when you are instructing a computer—you need to know what commands it already knows so that you can stick to the shared vocabulary.

Some commands stand alone, such as

`clap`

Others use *parameters*, creating, roughly, a related collection of commands. For example, if you know how to follow a command such as

`lift` right `foot across and` in front of `body`

you can also

`lift` left `foot across and` in front of `body`

or

`lift` right `foot across and` behind `body`

The family of commands is captured by

`lift` *x* `foot across and` *y* `body`

where the parameter *x* can be filled in with `right` or `left` and the parameter *y* can be filled in with `in front of` or `behind`. This idea is used in many computing contexts as well.

Parameters can also be numeric, like the "2" in

`tap right heel` 2 `times in front`

Once the set of commands and their associated parameters is established, the commands can be strung together in *sequence*:

`lift` right `foot across and` in front of `body`
`touch` right `foot with` left `hand`
`put` right `foot on ground`

The ordering of the commands is important—they are meant to be followed in standard reading order—top to bottom, left to right—at least in English. The command sequence above has you touch your foot while it's lifted up and already pretty close to your hand. In contrast, reordering the same commands like this:

`touch` right `foot with` left `hand`
`put` right `foot on ground`
`lift` right `foot across and` in front of `body`

leaves you standing on one leg. Same commands, different order, different result.

If you put the commands we've covered together in the right order and with the right parameter values, you are about 80 percent of the way to dancing to "Cotton Eye Joe" by Rednex. You're welcome. I'm glad attending Sports Camp Judaea could pay off for both of us.

The notion of writing down commands, some with parameters, in sequence is how we convey the most basic tasks to people, especially for performances like dances (the order in which to do the steps) or songs (the order in which to sing the notes) or theater (the order in which to say the lines). Conveying more complex tasks combines command sequences with the elements we'll discuss in later chapters, opening up the door for car repair ("replace the spark plug when . . ."), cooking ("roll out the pie crust until . . ."), and, yes, even sports ("steal second on a pitcher who . . .").

Commands are also quite handy for conveying tasks to machines. Let's look at a few examples that you can experiment with, starting with programming universal remotes.

MACRO MAY I?

In chapter 1 and chapter 2, we talked about the distinction between steering the Mars rover by remote control or by giving it commands to follow. Although both involve telling a machine what to do, the former requires your conscious attention in deciding which command to deliver when ("Turn left now, go straight now, turn right now"), while the latter has to include all the relevant information so that the machine can carry out the entire behavior without that kind of moment-to-moment oversight. It's very much like the distinction between square dancing and the Cotton Eye Joe dance. From the point of view of the dancer, square dancing is like being the rover controlled remotely by someone at NASA calling out each instruction. But you can do the Cotton Eye Joe dance without external guidance. For the person conveying the tasks, the distinction is between calling out square dancing commands on the fly and writing down the choreography of a line dance for people to learn on their own— analogous to the difference between *controlling* a computer and *programming* one, as we discussed in the previous chapter.

Since a remote control is the poster child for what this book is *not* about, it may be a bit surprising that I'm using it as a positive example.

But that's because some remotes can accept a special kind of program called a *macro*.

The term macro in computing is short for macroinstruction, at least originally. The idea was developed in the early days of computing when programs were written in what is known as *machine language*. Machine language, or sometimes assembly language, was, as the name implies, a language for telling machines what to do. But doesn't that describe *all* computer languages? Well, yes. But the commands in these primitive languages were directly implemented in the physical hardware of the machine, telling the circuits how to shuffle data around in the computer's memory. Modern computer languages let you express commands at a higher level of abstraction that leaves a little space between you and the computer. Writing computer programs in machine language is like describing an entire line dance as a sequence of descriptions of precisely which groups of muscles need to contract, with what strength, over what period of time. It's the sort of painstaking attention to detail that would make you want to

```
contract your supraspinatus at 10% for 100 ms
contract your infraspinatus and teres minor at 30% for 80 ms
contract your subscapularis at 60% for 200 ms
contract your lateral deltoid at 20% for 750 ms
```

(that is, *throw up your hands*) in frustration.

Early programmers created macroinstructions as a form of shorthand. They could set things up so that writing "ThrowUpYourHands" in a program (something easy for the programmer to remember) would automatically expand into the steps above (something easy for the computer to execute).

These days, the shortened form "macro" is used to refer to predefined command sequences that can be run with a single command—you might call them shortcuts—and that are available to people to use in contexts far from coding in machine language. For example, a universal remote control could let you create a macro for sending out the commands to configure a home theater system to play a DVD without needing to push buttons on two different remotes:

```
TV: power
Blu-ray Player: power
```

```
TV: input select to Blu-ray Player
Blue-ray Player: open disc tray
```

That would turn on the TV and the DVD player, tell the TV to display the output of the DVD player, and open the DVD player disc slot so that you could put in the movie. A macro would trigger all of that with a single press.

We can see in this command sequence the three main concepts described earlier. There are *commands* that can be carried out by the home-entertainment devices. There are *parameters* for some of these commands, such as that the TV input should be set to the Blu-Ray player rather than to, say, your computer. And the choice of *sequence* matters: you want the TV to be on before you set its input because otherwise it'll just ignore you. In fact, sometimes it's necessary to introduce delays in the sequence (such as a command to wait for two seconds after turning on the DVD player before issuing the next command) so that the devices are ready to hear what comes next.

A macro can do all that for you. Macros save you from needing to remember the sequence or to press all those buttons in the right sequence with the right delays.

Or, if the idea of remote control macros doesn't resonate with you, consider the benefits of a contact list on your phone. Do you remember having to memorize a different sequence of button pushes for everyone you wanted to call, like a caveperson? You can think of the contact list as a library of macros that relieves you of this burden.

How do you tell the remote that you want to create this macro? That's another nice example of a command sequence. On a typical universal remote, it would look like:

```
click "define macro"
select the button that will trigger the macro in the future
click "add sequence"
push the series of buttons on the remote that make up the macro
click "done"
click "update remote"
```

Is it worth memorizing a six-step sequence to avoid having to memorize a four-step sequence? Maybe not for you. But if you live with people

Figure 3.1 Early humans used a primitive interface to communicate.

who find it easier to use a one-press macro than to remember the steps for setting up the DVD, it could be extremely helpful.

UNDER THE MACROSCOPE

More useful, at least in my daily life, is a macro-based text editor called Emacs I've been using since my college days. Text editors are how we compose and modify text documents. They are a lot like word processors, but text editors usually just focus on the basics in terms of producing unformatted files without extra information such as font changes or headers and footers. I wrote this sentence in a text editor, for example. Then I pasted it into Google Docs, where I'm doing all my writing. I have Google Docs set up to use the same keyboard commands I normally use in Emacs, but preparing it in Google Docs makes the end result easier for me to share with my editor, Gita Manaktala, at the MIT Press. Now that I think about it, though, since Google Docs lets me save files in text-only ".txt" format, one could argue that it can function as a text editor, too. Regardless, I want to tell you about Emacs because it illustrates the power of macros in a form that you might find useful for your own purposes.

For some reason, one's choice of preferred text editor is a deeply personal decision that elicits spirited discussions about how anyone could choose otherwise, not unlike a political candidate, the greatest rock guitarist, or the proper size of the chips in chocolate chip ice cream. I'm a devotee of Emacs, which is short for "Editor Macros." The VH1 to Emacs' MTV in the "editor wars" is vi (pronounced vee eye, short for "visual" because, yes, early text editors were so primitive that you had to edit your text without being able to see what you were doing and making editing visual was a step forward!). I consider myself to be a pretty open-minded, let-a-thousand-flowers-bloom kind of guy, but my commitment to Emacs is strong. I once proclaimed "Emacs is my life!," much to my wife's immediate and visceral shock. (I'm guessing she's a vi user.)

One of the things that makes Emacs so dreamy, um, I mean "special," is that all the stuff that text editors do (move the cursor up one line, jump back to the beginning of the file, save a file, do the hokey pokey and turn itself about, etc.) are all commands in a built-in language called Emacs LISP. Which commands are associated with which keypresses is all under the user's programmatic control. For example, typically, "?" is assigned to the command

`insert` a question mark `into the document at the cursor position`

But in Emacs it need not be. In some contexts, it'd be better for the question mark key to bring up some help text, for example.

With some practice, you can redefine the behavior of Emacs to fit your own needs, freeing you from blind dependence on other people's code. In fact, back in the early 1990s, I was working as a research assistant to Tom Landauer, an influential cognitive scientist and author of *The Trouble with Computers*. I spent a chunk of my time helping to extend Emacs's ability to display a variety of calendars, edit specification files for a neural network simulator, interpret email attachments, and even generate text-based "magic eye" 3D pictures. Looking back, we sure did have a lot of fun together . . .

Even though (some of) what I was doing was intended to make Landauer and his research group more productive, he refused to use Emacs. As a usability expert, he had a knack for putting himself into a novice mindset and then getting lost there. He found Emacs confusing and frustrating. To jump the cursor to the end of a file in Emacs, you hold down the "meta" key (often associated with "option" or "escape" on the keyboard) and type a greater-than sign, abbreviated `Meta->`. This keypress involves pushing three buttons simultaneously—meta, shift, and the key

```
v              v              v              v              v
How do I feel aHow do I feel aHow do I feel aHow do I feel aHo
bout Emacs? Welbout Emacs? Welbout Emacs? Welbout Emacs? Welbo
l, that's a coml, that's aa col, that's  aa ol, that's  aa ol,
plex question tplex quesstion tpex queesstion pex queesstion p
hat should be ehat shoould be ehatshhoould be ehtshhoould be e
xamined from alxaminned from alxaminned from alxamnned from al
l possible angll posssible angll posssible angll psssible angl
es. On the one es. OOn the one es. OOn the one es.OOn the one
hand, Emacs is hand,, Emacs is hand,, Emacs is han,, Emacs is
just a piece ofjust  a piece ofjust  a piece ofjus  a piece of
 software, inca softwaare, inca softwaare, inca oftwaare, inca
pable of reciprpable oof reciprpable oof reciprpble oof recipr
ocating complexocating ccomplexocating ccompleocating ccompleo
 human emotions human emottions human emottins human emottins
. On the other . On the otheer . On the oteer . On the oteer .
hand, I would bhand, I would bbhand, I wuld bbhand, I wuld bbh
e completely loe completely loe  compltely loe  compltely loe
st without beinst without beinst wwihout beinst wwihout beinst
g able to use Eg able to use Eg able to use Eg able to use Eg
macs on a dailymacs on a dailymacs on a dailymacs on a dailyma
 basis. So, I g basis. So, I g basis. So, I g basis. So, I g b
^              ^              ^              ^              ^
```

Figure 3.2 By blurring your vision so that the v's merge, you can see a 3D image, generated in Emacs, that answers the question of how the author feels about Emacs.

that has the greater-than symbol. To save the current file to disk, you hold down the control key and type "x," then type "s" for "save," abbreviated "Ctrl-x s." Landauer disdainfully called these key sequences "chords" and said he preferred to stick to piano keyboards when playing chords. I learned a lot from him about how to make computers more usable, but I did not find his critique of Emacs convincing. I just wish he could have seen Emacs through my heart-shaped eyes. Then he would have understood. Sigh.

Because all of Emacs's functions are commands, everything you type into the editor can be viewed as a command sequence. Typically these commands are fed to Emacs one at a time, square dancing style, with the person typing as the caller and Emacs responding as the dancer. I push "t e h Ctrl-t" and Emacs inserts "t," "e," "h," and then transposes the last two characters to make "the." But you can also ask Emacs to start remembering what you type. If I were to hit "Ctrl-x (", then type

"All work and no play makes Jack a dull boy.", and then press "Ctrl-x)", I would have defined a command sequence, specifically a *keyboard macro*. Hitting "Ctrl-x e" *executes* this macro, typing out the sentence about Jack again. Each "C-x e" types it out again. You can do it over and over again, all day, like Jack Nicholson's character in *The Shining*. But with less typing!

Keyboard macros are particularly handy for rearranging existing text in some simple but tedious way. For example, if I want to reverse the lines of a file so that it reads bottom to top in a new file, I can cut and paste the lines one at a time, moving each one from the top of the file to the top of the new file, like unstacking and restacking a set of dishes. The command sequence would be:

mark the current cursor position
go down one line
cut everything between the mark and the new cursor position
switch to the other file
go to the top of the file
paste
switch to the other file

Or, in terms of Emacs keystrokes:

Ctrl-space Ctrl-n Ctrl-w Ctrl-x o Meta-< Ctrl-y Ctrl-x o.

Putting this sequence between "Ctrl-x (" and "Ctrl-x)" would make it a keyboard macro. I can do "Ctrl-x e" a few times to see that it's working. Then "Meta-0 Ctrl-x e" says "just keep doing it until something goes wrong" (specifically, we've run out of lines to copy). Ta da!

Oh, wait, that actually didn't work. Some of the lines got broken up. Ugh, okay, I see what went wrong. Typing the commands one at a time

and turn myself about
hake,
I give my right hand, shake, shake, sI put my right hand out
I put my right hand in

Figure 3.3 The "Ugly Mug," from 1872, was likely an inspiration for the "Hokey Pokey" song. A buggy keyboard macro mangles the lines of its first verse instead of reversing them.

would make it clear that this macro doesn't handle long lines properly. In particular, the "Ctrl-n" command doesn't go to the next line in the file, it goes to the next line on the *screen* (partway through the line of text if the line of text is wider than the screen). After a bunch of undo commands (Ctrl-x u), it works to replace "go down one line" with the combination of "go to the end of the current line" (Ctrl-e) and "Go forward one character" (Ctrl-f). This revised keyboard macro does the job.

Whew! Okay, this example illustrates a key concern that arises when we switch from interactively issuing commands—control—to creating and running command sequences—programs. When things go wrong with a program, we may not be there to see it. A machine might continue to carry out our literal commands even after things have gone badly off track. Debugging our instructions is a constant concern when conveying tasks to machines, something we'll return to throughout the book.

The development of Emacs began in the 1970s, which is ancient history in computing and also in feathered hairstyles. But the idea of controlling machines with command sequences goes back a lot further. Almost two thousand years ago the Greek engineer Hero of Alexandria designed a theatrical delivery cart that he could program with command sequences. Using weights to pull a rope wound around an axle, the two independent drive wheels of the cart could accept any of three commands: forward, backward, pause. Coordinating the sequence of commands allowed the cart to trundle forward or backward, spin right or left, or hold patiently in place so that the audience could get a good look at it. The forward and backward commands may have even supported separate speed parameters for individual wheels, which would have allowed the cart to move in circular arcs as well.

The cart reminds me of a toy truck I coveted back in the 1970s called Big Trak. It accepted a sequence of up to sixteen commands, including all the movement commands just described, but also commands for dumping out the contents of its trailer or firing a laser. So cool! Of course, I'm now too old and dignified to . . . hold on, I'm going to go see if I can find it on craigslist.

No luck. Anyway, the surviving documentation for the delivery cart describes how the cart could be programmed by demonstration (aka example instructions from our 2 × 2 grid): rolling it backward through its routine, the movements would be "stored" in the winding of the ropes. It was a lot like how we define keyboard macros. Computer scientist Noel Sharkey at the University of Sheffield thinks programmable machines may have been in use even centuries earlier, but Hero is the most ancient programmer whose macro designs are known today.

QUESTIONS EXTRAORDINAIRE

In junior high, I had a reputation for putting together surveys and going around the cafeteria during lunch getting people to answer my questions. I mostly made the questions about the other students and their likes/dislikes, as that's information even junior high school students enjoyed sharing. I'd tally up their answers at the end of the week and then share the results the following Monday. It was a lot of work to visit all the various tables and to count up the results at the end. But it was worth it because it gave me a chance to have friendly conversations with the full cross section of students at my school. Socializing didn't come naturally to me.

These days, my department uses online questionnaires all the time to gather information from students and faculty. The most recent was a PhD recruitment day attendance poll intended to help us figure out who would be around to talk to the newly accepted PhD students. Wouldn't it be sad to have a meet-the-new-students day if only the new students showed up? We used Google Forms to build an online form that posed a series of questions:

Name—short answer text box
Role—multiple choice of student versus faculty
Preferences for faculty presentations—multiple choice grid for in-person,
    prerecorded, or Zoom, with choices of preferred, willing, unwilling
Additional thoughts—paragraph answer text box

The result is a kind of program that we call a questionnaire, one that you might find useful to experiment with in any of several available systems—Google Forms and Qualtrics are two I've used, but searching "online form builder apps" reveals several more you can try. The giveaway that the questionnaire you create is a program is that when you create one of these surveys, there's a little preview button that lets you try out what you created. The fact that there is a distinction between *editing* the questionnaire script and *running* it means that the survey you're building is a command sequence to be executed by a computer. In fact, the three main attributes of command sequences I listed at the top of the chapter are apparent in questionnaire construction: commands, sequences, and parameters.

First, as the constructor of the questionnaire, I get to command the form-maker app to insert checkboxes, text boxes, and so forth. Each of these elements of a questionnaire will show up with programmatic elements: multiple-choice checkboxes have the option of preventing a user from ticking more than one per question, a phone number entry box can be configured to reject entries that don't have the right number of digits, and so forth.

Second, I can *sequence* these commands however I want on the page, controlling the order in which the user will fill them in.

Third, the commands that create the elements of a form can have *parameters*. For example, the text that precedes a checkbox is a parameter of that checkbox, as would be the decision that a particular text entry box should limit the number of characters that can be typed into it, or that numeric entries can only be within some specified range. All of these options have an impact on how the questionnaire is presented to the people filling it out and what it enables them to do.

I chose questionnaires as the first try-it-yourself example in the book because they are quite simple: ask a question, get an answer, ask a question, get an answer, and so on. But the truth of the matter is that you have a rich set of options available to you when you construct a questionnaire. You can have it skip questions depending on previous answers (see chapter 4), validate the answers people give (chapter 5), re-ask questions multiple times (chapter 6), and name blocks of questions to reuse them in other questionnaires (chapter 7). Regardless of the specific application, all the various features for telling machines what to do work together to let you express the behavior you want. As you become more fluent in their use, you'll start to notice them in all sorts of computer interfaces and you won't like it when some are available only in a constrained form.

By the way, that reminds me to ask you if you think the cafeteria's tater tots or french fries are better. I'll let you know what everybody else thinks on Monday.

"GIMME A BLUE!"

Card stunts, in which a stadium audience holds up colored signs to make a giant, temporary billboard, are like flash mobs where the participants don't need any special skills and don't even have to practice ahead of time. All they have to do is show up and follow instructions in the form of a short command sequence. The instructions guide a stadium audience to hold aloft the right poster-sized colored cards at the right time as announced by a stunt leader. A typical set of card-stunt instructions begins with instructions for following the instructions:

<u>listen to instructions carefully</u>
<u>hold top of card at eye level</u> (not over your head)
<u>hold indicated color toward field</u> (not facing you)
<u>pass cards to aisle on completion of stunts</u> (do not rip up the cards)

Figure 3.4 Virginia Tech fans can use card stunts to spell out supportive messages like "Let's Go." *Source:* Cal Sport Media / Alamy Stock Photo. Photographer: Scott Taetsch.

These instructions may sound obvious, but not stating them surely leads to disaster. Even so, you know there's gotta be a smart alec who asks afterward, "Sorry, what was that first one again?" It's definitely what I'd do.

Then comes the main event, which, for one specific person in the crowd, could be the command sequence:

1. Blue
2. Blue
3. Blue

Breathtaking, no? Well, maybe you have to see the bigger picture. The whole idea of card stunts leverages the fact that the members of a stadium crowd sit in seats arranged in a grid. By holding up colored rectangular sign boards, they transform themselves into something like a big computer display screen. Each participant acts as a single picture element— person pixels! Shifts in which cards are being held up change the image or maybe even cause it to morph like a larger-than-life animated gif.

Card stunts began as a crowd-participation activity at college sports in the 1920s. They became much less popular in the 1970s when it was generally agreed that everyone should do their own thing, man. In the 1950s, though, there was a real hunger to create ever more elaborate displays. Cheer squads would design the stunts by hand, then prepare individual instructions for each of a thousand seats. You've got to really love your team to dedicate that kind of energy. A few schools in the 1960s thought that those newfangled computer things might be helpful for taking some of the drudgery out of instruction preparation and they designed programs to turn sequences of hand-drawn images into individualized instructions for each of the participants. With the help of computers, people could produce much richer individualized sequences for each person pixel that said when to lift a card, what color to lift, and when to put it down or change to another card. So, whereas the questionnaire example from the previous section was about people making command sequences for the computer to follow, this example is about the computer making command sequences for *people* to follow. And computer support for automating the process of creating command sequences makes it possible to create more elaborate stunts. That resulted in a participant's sequence of commands looking like:

up on 001 white
   003 blue
   005 white
   006 red
   008 white
   013 blue
   015 white
   021 down
up on 022 white
   035 down
up on 036 white
   043 blue
   044 down
up on 045 white
   057 metallic red
   070 down

Okay, it's still not as fun to read the instructions as to see the final product—in this actual example, it's part of an animated Stanford "S." To execute these commands in synchronized fashion, an announcer in the stadium calls out the step number ("Forty-one!") and each participant can tell from his or her instructions what to do ("I'm still holding up the white card I lifted on 36, but I'm getting ready to swap it for a blue card when the count hits 43").

As I said, it's not that complicated for people to be part of a card stunt, but it's a pretty cool example of creating and following command sequences where the computer tells us what to do instead of the other way around. And, as easy as it might be, sometimes things still go wrong. At the 2016 Democratic National Convention, Hillary Clinton's supporters planned an arenawide card stunt. Although it was intended to be a patriotic display of unity, some attendees didn't want to participate. The result was an unreadable mess that, depressingly, was supposed to spell out "Stronger Together."

These days, computers make it a simple matter to turn a photograph into instructions about which colors to hold up where. Essentially, any digitized image is already a set of instructions for what mixture of red, blue, and green to display at each picture position. One interesting challenge in translating an image into card-stunt instructions is that typical images consist of millions of colored dots (megapixels), whereas a card-stunt section of a stadium has maybe a thousand seats. Instead of asking each person to hold up a thousand tiny cards, it makes more sense to compute an average of the colors in that part of the image. Then, from the collection of available colors (say, the classic sixty-four Crayola options), the computer just picks the closest one to the average.

If you think about it, it's not obvious how a computer can average colors. You could mix green and yellow and decide that the result looks like the spring green crayon, but how do you teach a machine to do that? Let's look at this question a little more deeply. It'll help you get a sense of how computers can help us instruct them better. Plus, it will be our entry into the exciting world of machine learning.

There are actually many, many ways to average colors. A simple one is to take advantage of the fact that each dot of color in an image file is stored as the amount of red, green, and blue color in it. Each component color is represented as a whole number between 0 and 255, where 255 was chosen because it's the largest value you can make with eight binary digits, or bits. Using quantities of red-blue-green works well because the color receptors in the human eye translate real-world colors into this same representation. That is, even though purple corresponds to a specific wavelength of light, our eyes see it as a particular blend of green, blue, and red. Show someone that same blend, and they'll see purple. So, to summarize a big group of pixels, just average the amount of blue in those pixels, the amount of red in those pixels, and the amount of green in those pixels. That basically works. Now, it turns out, for a combination of physical, perceptual, and engineering reasons, you get better results by *squaring* the values before averaging, and square rooting the values after averaging. But that's not important right now. The important thing is that there is a mechanical way to average a bunch of colored dots to get a single dot whose color summarizes the group.

Once that average color is produced, the computer needs a way of finding the closest color to the cards we have available. Is that more of a burnt sienna or a red-orange? A typical (if imperfect) way to approximate how similar two colors are using their red-blue-green values is what's known as the Euclidean distance formula. Here's what that looks like as a command sequence:

```
take the difference between the amount of red in the two colors
square it
take the difference between the amount of blue in the two colors
square it
take the difference between the amount of green in the two colors
square it
add the three squares together
take the square root
```

So to figure out what card should be held up to best capture the average of the colors in the corresponding part of the image, just figure out which of the available colors (blue, yellow green, apricot, timberwolf, mahogany, periwinkle, etc.) has the smallest distance to that average color at that location. That's the color of the card that should be given to the pixel person sitting in that spot in the grid.

The similarity between this distance calculation and the color-averaging operation is, I'm pretty sure, just a coincidence. Sometimes a square root is just a square root.

Stepping back, we can use these operations—color averaging and finding the closest color to the average—to get a computer to help us construct the command sequence for a card stunt. The computer takes as input a target image, a seating chart, and a set of available color cards, and then creates a map of which card should be held up in each seat to best reproduce the image. In this example, the computer mostly handles bookkeeping and doesn't have much to do in terms of decision-making beyond the selection of the closest color. But the upshot here is that the computer is taking over some of the effort of writing command sequences. We've gone from having to select every command for every person pixel at every moment in the card stunt to selecting images and having the computer generate the necessary commands.

This shift in perspective opens up the possibility of turning over more control of the command-sequence generation process to the machine. In terms of our 2 × 2 grid from chapter 1, we can move from telling (providing explicit instructions) to explaining (providing explicit incentives). For example, there is a variation of this color selection problem that is a lot harder and gives the computer more interesting work to do. Imagine that we could print up cards of any color we needed but our print shop insists that we order the cards in bulk. They can only provide us with eight different card colors, but we can choose any colors we want to make up that eight. (Eight is the number of different values we can make with 3 bits—bits come up a lot in computing.) So we could choose blue, green, blue-green, blue-violet, cerulean, indigo, cadet blue, and sky blue, and render a beautiful ocean wave in eight shades of blue. Great!

But then there would be no red or yellow to make other pictures. Limiting the color palette to eight may sound like a bizarre constraint, but it turns out that early computer monitors worked exactly like that. They could display any of millions of colors, but only eight distinct ones on the screen at any one time.

With this constraint in mind, rendering an image in colored cards becomes a lot trickier. Not only do you have to decide which color from our set of color options to make each card, just as before, but you have to pick which eight colors will constitute that set of color options. If we're making a face, a variety of skin tones will be much more useful than distinctions among shades of green or blue. How do we go from a list of the colors we wish we could use because they are in the target image to the much shorter list of colors that will make up our set of color options?

Machine learning, and specifically an approach known as clustering or unsupervised learning, can solve this color-choice problem for us. I will tell you how. But first let's delve into a related problem that comes from turning a face into a jigsaw puzzle. As in the card-stunt example, we're going to have the computer design a sequence of commands for rendering a picture. But there's a twist—the puzzle pieces available for constructing the picture are fixed in advance. Similar to the dance-step example, it will use the same set of commands and consider which sequence produces the desired image.

MAKE ME A PUZZLE

Ji Ga Zo is a three-hundred-piece jigsaw puzzle, once quite popular in Japan and sold in the US by Hasbro. The word *jigazō* means "self-portrait" in Japanese, but it also sounds like how you might transliterate "jigsaw" into Japanese—similar to how "mele kalikimaka" is "Merry Christmas" in Hawaiian. At three hundred pieces, it doesn't sound like a particularly noteworthy puzzle—that's typical for a puzzle rated for eight-year-olds. What makes Ji Ga Zo special is that it can be reconfigured to depict your face. Or anyone's, all with exactly the same set of pieces. If you feed a selfie into the Ji Ga Zo software, it will spit out a map of how to arrange the three hundred completely interchangeable pieces into a low-res but quite recognizable sepia rendering of the picture you gave it.

The mosaic principle Ji Ga Zo uses is closely related to the card-stunt idea: break a big picture into lots of smaller pieces so that when they are viewed all together, we see the original picture again, although at lower resolution. The big difference, though, is that Ji Ga Zo uses a fixed collection of three hundred pieces. It's as if every card stunt used the same cards but, to make a particular message, the cards had to be reshuffled among the fans and held up in a different arrangement. It's a neat trick. And trick is a reasonable word for it—Ji Ga Zo was created through a collaboration among a software expert, a computer graphics pioneer, and a magician. I'm guessing they met while walking into a bar.

Here's how it works. Each Ji Ga Zo piece has two sides. The front side has the surface that will become part of the picture. Each piece varies in its shade and its gradient—how the shade changes across that piece. On the other side is a glyph—a little symbol that uniquely identifies the piece. The glyphs are eclectic: for example, a pretzel on a pink background, headphones on a blue background, a poodle on a green background, and, somewhat meta-ly, a picture of a puzzle piece on a yellow background. The choice of symbols themselves is irrelevant except that they are all different and each one has the property that you can tell which of the four orientations it is in so the piece will be facing the right way when the puzzle is assembled. For example, there is a heart, a club, and a spade on a yellow background, but no diamond. After all, a diamond is the same upside down and therefore its orientation is ambiguous.

The arrangement of the symbols constitutes the map that the puzzle solver needs to follow to arrange the pieces and make the face. I had one of these puzzles a decade or so ago and it was super fun to take a picture of a friend, then present that person with the map so they could assemble the puzzle, only for them to discover that they just made a picture of themselves!

Although it can take a while to assemble all of the pieces, it's still a pretty rote process—sift through the pile of pieces looking for, say, the butterfly net on the blue background and click it into the right place in the right orientation. Repeat 299 more times according to the map. The real heavy lifting here is left to the computer, which has to figure out how to arrange the palette of three hundred pieces so that the desired picture is formed.

A hint about how this feat is accomplished can be found when running the Ji Ga Zo software. When you feed it the target image, it displays that image, the map, and what the finished puzzle will look like. But it's not that great at first. Over a period of about a minute, the pieces on the map swap themselves around and you can see the image go from vaguely the right shape to an accurate rendition of the target image. It's fun to watch, but it also suggests that the software is iteratively improving the layout to create a more and more accurate picture. That is, the software has a built-in sense of how good the puzzle looks at the moment, and considers how to tweak the arrangement of the pieces to make the picture even better. When it runs out of ways to improve the picture, it stops.

This iterative approach would be overkill when constructing mosaics for card stunts. That's because the choice of card at each position in the picture is a location-specific choice. We use a red card for row 46, seat 5, because that's what works best at that position in the picture. But think of what would happen if we did the same for Ji Ga Zo. The best puzzle piece for row 10, column G, and the best puzzle piece for row 18, column D, might be the same piece. You can't use the same piece in two places because, well, physics; these are real pieces, not just bits. We could have the program pick the best piece from those that haven't yet been assigned to some other position, but that's likely to yield only so-so results. In fact, that's probably close to what the Ji Ga Zo software uses as its starting point. But there are better arrangements than what you get from this greedy "you can't have that piece because it belongs to me now" approach. The computer considers lots of possible trades and rearrangements and slowly works its way toward better and better solutions to the problem of using the given three hundred pieces to make the given

Figure 3.5 Left: You can make a picture of a baseball glove out of a 15 × 20 array of image patches (each of the three hundred image patches is 40 × 40 pixels). Right: Those exact same three hundred image patches can be rearranged to look like a well-known athlete.

portrait. Viewing each puzzle piece as a command, it's working out the sequence that comes closest to creating the desired portrait. It's doing a kind of automatic programming—an example of a much more general idea we'll look at next.

THAT'S YOUR LOSS

If we had written the Ji Ga Zo program, we'd be right to be pretty proud. But let's say that we want to be even prouder by teaching a computer to code up a program on its own. We still need to provide some guidance so that the computer knows where we want it to go. We're going to get a little bit conceptual for a moment, so stay with me. We need to tell the computer three things so that it can go off and do its work and come back with a solution:

- Which programs should the computer consider?
- How should the computer *score* programs to determine which ones are better than others?
- What procedure should the computer use to sift through the programs under consideration to find the one that scores best?

In computer science jargon, the programs under consideration are the *representational space*, the method of scoring is the *loss function*, and the procedure for finding the best one is the *optimizer*. You don't really need to know these terms, but dropping them into casual conversation will make you look pretty cool at the nerd square dance.

In the context of Ji Ga Zo, the "programs" under consideration are the arrangements of the puzzle pieces, a simple kind of command sequence that the computer will give to you to guide you to build an image out of puzzle pieces. The representational space, therefore, is all of the fifteen-by-twenty arrangements of the three hundred pieces, where each piece gets a location in the grid and a rotation to appear in.

To score one of these programs for Ji Ga Zo, the computer looks at every position in the image and asks, "How different is the shading at that position in the target image from the shading at the corresponding part of the current puzzle image?" If the puzzle image is dark where the target image is light, that's bad. If the bright and dark regions line up, that's good.

The same idea we talked about when computing color distances can be used here as well. Take the difference in brightness between the target image and the puzzle image at each position, square it, and sum their squares over the entire image. If the puzzle image is an exact match for the target image, this value will be zero. If the light and dark parts of the target image tend to line up with light and dark parts of the puzzle image, the value will be small. If the puzzle image looks like a photographic negative of the real image, the value will be huge. The bigger the difference between the original image and the computer's current arrangement of the pieces, the bigger (the worse) the score. In short, your loss function returns a large value because you're losing the game.

At this point, the computer can take any arrangement of pieces in the representational space and score it with the loss function. In principle, all it would need to do is to try all of them and choose the arrangement with the lowest loss with respect to the target image. Problem solved! In practice, however, there are too many arrangements to check. In the case of Ji Ga Zo, there are a mind-crushing 10^{795} arrangements in it. That's way bigger than the number of particles in the universe. In fact, it's way bigger than the number you would get if each particle in the known universe had, in its pocket, a contact list with the names of all the other particles in the known universe. (You know, in case they want to hang out sometime.)

Since we can't check how close every possible arrangement of the puzzle pieces is to the target image (in jargon: we can't calculate the loss of all arrangements in our representational space), we need a way to consider only the most promising arrangements. For some loss functions (like the one we used in the picture-to-card-stunt problem proposed in the previous section), there are specialized approaches that run fast and are guaranteed to produce the best answer. For others, like the Ji Ga Zo problem, making a good initial guess and then considering small-scale rearrangements may be the best you can do. But perhaps that's fine. After all, Ji Ga Zo succeeds if your friends say "Oh, cool!" when they see their faces emerge from the pieces. It doesn't have to be the absolutely best possible version of the image for that to happen. Cool is cool.

MACHINES LEARNING TO LEARN

The components we just discussed—describing a set of arrangements to consider, a way to score them, and a way to sift through the set to find one that's good (enough)—constitute a three-ingredient *machine learning recipe*. It's a great way of delegating to a computer the process of coming up with the programs it needs to accomplish your objectives. Of course, there's no free lunch. You still have to decide on the ingredients. And just as sequences of commands that you write can have bugs, your choice of ingredients can lead the machine to erroneous results, too.

The loss function, in particular, is notoriously challenging to get right. It should capture your sense of what's a good answer and what's a bad answer in a formal mathematical way. It's not enough to just say "I know a good answer when I see it" because you're asking the computer to do the search for you. To be fair, it's a hard problem to solve in everyday life as well. My favorite example is the government agency that wanted to rid a community of cobras. They defined a loss function they thought would do the trick—pay a reward for every dead cobra brought to them. But instead of creating an incentive for people to kill the cobras, the result was that community members began breeding cobras for the reward money. Oops! That's pretty much the opposite of what they wanted and it might have been possible to avoid had the agency designed the loss function better. The whole concept of tax loopholes is about finding ways to game a loss function. An example from computer science is the software that was shown a stack of colored blocks and was told to make a second stack with the same color sequence as the first stack. It made a stack of the right height and then painted all the blocks black. Black matches everything.

Despite the challenges in picking the right loss function, machine learning is an awesomely powerful mechanism for getting machines to help us out. Here are a couple of examples you might recognize that use the machine learning recipe. Web search engines (Google, Bing, Duck-DuckGo) can be thought of this way. Their representational space is the collection of web pages they have gathered from the internet. The loss function is a measure of relevance of a specific page to your most recent query (and sometimes even to other elements of your browsing history). They find low-loss pages using their home-grown index that maps the words in your query to pages likely to score well. The search engine then brings together the three components—from the collection of indexed web pages (within its representational space), it displays the best pages (uses the optimizer to identify pages) with respect to their appropriate-ness to the user's query (with low loss).

Driving direction apps follow the recipe as well. The representational space is the set of all paths between the two locations you provide. The loss function is typically some measure of the desirability of the roads on that path—how well traffic is moving and the overall distance, for example. Specialized software optimizes the routes by this measure in the blink of an eye. Many map apps will let you tweak the loss function and the representational space in limited ways. For example, you can tell the app to avoid highways or allow dirt roads, changing the representational space. You can also select whether you want the loss function to be based on speed or distance, where the former may end up suggesting a path that uses roads that are faster even if they are less direct. Or you might specify that you want the route with the lowest environmental impact—fewer starts and stops, fewer hills to climb, and so forth.

With these ideas in mind, we can attack the color-choice problem from earlier in the chapter: if limited to eight colors, select the best eight to represent an image in a card stunt. The representational space can be all possible choices of eight colors. To measure the loss, we can say that we want the distance from each color in the image to the closest color in the computer's set of eight to be small, say, by adding up the total distances over all the pixels in the image. In practice, we'd probably use the squared distances we mentioned earlier, in part because there are great methods available to minimize this loss. (Hmmm, between the style of dancing, the shape of the colored cards, mathematical self-multiplication, and a not very athletic camper, we could say that this chapter has been brought to you by "squares.")

This approach is an example of *unsupervised learning*, one of the most nebulous branches of machine learning in that there's not a lot of agreement on what it means. At a high level, though, an unsupervised learning problem is one that involves a collection of data with no particular external guidance as to what to do with it. In our color-choice problem, the data are the colors of the thousands or millions of pixels in a given image. An unsupervised learning algorithm would look for a manageable set of colors that would provide each pixel with a color close enough to look right. The machine learning software is in effect grouping or *clustering* the wide array of colors in the original into a much smaller set of representative colors. The same kind of approach can be used by, for example, streaming services to collect movies into fine-grained genre categories to help with recommendations—if you liked Disney's *Jungle Cruise*, maybe you'd enjoy one of these six other movies that feature Dwayne "The Rock" Johnson making his way through a jungle.

My Bellcore boss Tom Landauer led a team that applied an unsupervised learning method to words instead of colors or movies. They took a set of encyclopedia articles and characterized thousands of words by which articles they appeared in. Their version of unsupervised learning, called "latent semantic analysis," analyzed the patterns of how words

Figure 3.6 This still shows Dwayne Johnson in *Jungle Cruise*. Or possibly *Journey 2: The Mysterious Island*. No, no, it's *Rampage*. *Red Notice*? Oh, right, it's one of the *Jumanji* movies! *Source:* Entertainment Pictures / Alamy Stock Photo. © Columbia Pictures/Entertainment Pictures.

are used. It was able to figure out, for example, that words like "cat" and "feline" are highly related in terms of meaning and usage. So are "king" and "queen," and "pneumonia" and "lungs." Versions of this idea are in common use today to help retrieve relevant information from the web even when the words you choose for your query don't explicitly appear on the web pages they are related to: a search for "felines" will include web pages about cats on which the word "felines" never appears.

As we'll see in later chapters, the machine learning recipe is useful in many different settings beyond unsupervised learning and puzzle design. But let's look at one more way unsupervised learning can help choose a command sequence.

ONE FINAL WORDLE

Wordle, a word-puzzle variation of the classic Mastermind guessing game, became a global phenomenon in early 2022. The computer picks a hidden five-letter word and you get up to six guesses to find it. Each time you guess a word, the computer provides "hints," telling you which of the letters of your guess are in the right place, which are right but in the wrong position, and which are just plain wrong. My personal goal is to get the right answer in three guesses, which I've managed to do in about a third of the games I've played so far.

Using the machine learning recipe, maybe I can get some strategic help to create a command sequence that brings this probability up a bit. I'll formulate the problem this way: What pair of words (kind of like commands!) should I choose in my first two guesses to maximize the probability that I can solve the puzzle on my third guess? So, I'm interested in programs that are two-word sequences. My loss function is the chance that a word consistent with the hints from the first two will yield the answer word on the third guess. To search for the best such program, I turned to a very powerful and generalized technique called "my friend Justin." In addition to being super-smart, Justin had already written some programs to attack the Wordle puzzle. Also, we like to do things together.

So I texted Justin my challenge and, in under an hour, he had an optimizer up and running to find the best pair of words. We quickly realized that we needed to make some decisions to specify the problem more precisely. On the representational space side, there's a choice of what words to consider. The Wordle puzzle has two different word lists. One is a list of common five-letter words such as "plane" and "crest" that can serve as possible hidden words in answers. The other is a larger set of five-letter words that Wordle accepts as guesses, including obscurities such as "aalii" and "snirt." We decided to use the common words because it would be easier for the computer to search the smaller list and the solution would look nicer. However, the best sequence from the smaller common list might be strictly worse than what we'd get using the longer list. That's a sacrifice we decided to make for you, the reader. Before bringing in Justin to help, I'd been using the sequence laser-point, which I settled on because it uses lots of common letters and it sounds like something a tech-savvy professor might use in lecture. Justin informed me that `laser` is only on the longer, obscure list, so our search won't even consider it. Still, I hope it can find something better.

Next, there's the choice of loss function, which is subtle. After the two guesses are made, the resulting hints will be consistent with some possible answer words. One goal might be to make it so that, as often as possible, I want the set of possibilities at this stage to contain exactly one choice. That means that I'd be confident that my guess would have to be right. We called that the "unlucky guesser" loss function because it assumes that, whenever there is more than one word consistent with the hints, I'd get it wrong. The loss function accounts for that by striving to leave me no wrong choices whenever possible.

A reasonable alternative is one we called the "average Joe" loss function. It assumes that, from the set of words consistent with the hints, I'd pick one completely at random. So sometimes, even when there are multiple possibilities, I'd stumble onto the right one. The average Joe loss function encourages selecting two words that reduce the possibility space for the third guess as much as possible, on average. (We also considered the "lucky guesser" loss function but realized it would be completely uninformative. That's because it assumes I'd always get lucky and would pick the right answer from whatever choices remain, so it wouldn't matter what previous guesses I had made. Being optimistic may be good for the soul but it's unhelpful from a planning perspective.)

Is it better to assume an average Joe guesser or an unlucky guesser? I don't think it's obvious. Both are related to approaches commonly used in real engineering settings. The people who design control systems for planes or other safety-critical systems often adopt the perspective of the unlucky guesser to provide strong guarantees that uncommon situations won't result in fatal consequences. After all, "on average, you won't fly into a mountain" is true but a terrible advertising slogan for autopilots. On the other hand, designing solely for the unlucky guesser would leave an autopilot permanently grounded—there's always a scenario that would lead to a crash. We ultimately decided that the average Joe loss function was a good fit for the Wordle problem. It's worth keeping in mind, though, that choosing the loss function chooses the problem, and comparing problems is a lot more subjective than comparing solutions to a *given* problem. Computers may be great at finding the right answer, but we still need people to ask the right question.

After a few hours of computer time, Justin's program found that the best two-common-word sequence is `slant-price`, and it results in average Joe solving the Wordle in three guesses 46.3 percent of the time. Interestingly, it turns out that this sequence is also the best under the unlucky guesser loss function, leading to a single best answer 27.4 percent of the time. My choice of `laser-point` is only a little worse—43.1 percent (24.7 percent for unlucky guessers). So why am I only succeeding at getting the answer in three guesses 34.8 percent of the time? Well, it could just be statistical fluctuation. Also, I occasionally goof and guess a word that was ruled out by the hints on the first two words. But it could also be my own limitations, mainly not knowing which words are on the common list. That's another important thing to keep in mind about loss functions. They are almost always *proxies*—stand-ins for the thing you wish you could measure but can't. As powerful as the machine learning recipe is, it is built on subjective choices, and we have to be humble about the results that come out. Whenever people claim that the computer has "solved" a real-world problem, they are guaranteed to have made many assumptions about the real world that aren't true. Dig a little deeper before accepting the claim.

Thinking about Wordle a bit more, it's an unnecessary restriction to always make the same second guess. After all, the hint that comes back from the first guess provides extra information that could lead to a more informed guess and a much higher probability of success. That is, instead of the simple command sequence `slant-price`, we could choose commands *conditionally*. And that's precisely what we should talk about next.

4

SPLITTING ON CONDITIONALS
IF THAT'S WHAT YOU WANT!

"If" is a small but remarkably powerful word. Like a tiny quantum mechanical reality splitter, it creates two parallel worlds, the same except for one point of deviation that may snowball into larger differences. If I were a rich man, for example, the point of deviation would be my wealth and the changes that follow would come from me being wealthy. For one thing, all day long I'd biddy biddy bum. That's definitely not something that's happening in my current reality—I can biddy biddy bum for forty-five minutes tops, but only on the weekend. "If" creates two worlds, one real, one hypothetical.

The ability to consider alternative worlds is vital in the context of conveying tasks. It lets us tailor behavior separately to each of these worlds, representing a major step beyond the restricted fixed sequences of commands described in chapter 3. Jumping from *Fiddler on the Roof* to another Jewish reference, let's consider a siddur or prayer book. If you're not familiar with Jewish practices, don't worry; you'll catch on.

In Judaism, the day of the week and the phase of the moon both hold significance in the rhythms of life that are reflected in religious services. Instead of creating a separate book of prayers for each combination ("Okay, everyone grab the Wednesday waxing gibbous edition from under the seat in front of you and repeat after me . . ."), congregations use a single prayer book that includes instructions about which passages to read in specific circumstances. For example, a paragraph might be marked for use during a sabbath that falls on a holiday. Sometimes the inclusion is coarse-grained, such as reciting a psalm. Sometimes it's more fine-grained, such as adding individual words to a sentence. During the service, congregants evaluate the conditions described in the text and respond accordingly. Written as code, the text of a festival service might look like:

```
say "We give thanks for"
if sabbath (Friday to Saturday evening): say "Sabbaths for rest,"
say "appointed times for gladness, festivals and seasons for joy;"
if sabbath: say "this Sabbath Day, and"
say "this day of"
if Passover:
    say "The Feast of Unleavened Bread, the season of our Freedom."
if Shavuos (another holiday):
    say "The Feast of Weeks, the season of the Lawgiving."
```

So, on a sabbath during Passover, congregants would say:

We give thanks for Sabbaths for rest, appointed times for gladness, festivals and seasons for joy; this Sabbath Day, and this day of The Feast of Unleavened Bread, the season of our Freedom.

This one block of text contains four separate passages: one for a sabbath during Passover, one for nonsabbath days during Passover, one for a sabbath during Shavuos, and one for nonsabbath days during Shavuos. (Shavuos and Passover never fall on the same day; otherwise there would be seven separate possibilities.)

Here the conditions are relatively simple. But I found a siddur that included, I kid you not, these instructions:

Both these Prayers are omitted on New Moon, during the whole month of Nisan, on the thirty-third day of Counting the Omer, from the first day of Sivan until the second day after Pentecost, on the 9th and 15th of Ah, on the day before New Year, from the day before the Fast of Atonement until the second day after Tabernacles, on the Feast of Dedication, on the 15th of Shebat, on the two days of Purim, and on the two days of Purim Katon, the 14th and 15th of Adar Rishon. These prayers are also omitted in the house of a mourner during the week of mourning, and at the celebration of a circumcision.

Setting aside the question of why a passage needs such an elaborate set of conditions for omission and whether "celebration" is the right word to describe a circumcision, the use of conditional inclusion of passages allows one text to stand in for hundreds or even thousands of variants.

Effective use of *if* statements can change the course of history. In the build-up to the US Revolutionary War, colonists amassed weapons in

Concord, Massachusetts. Having caught wind of this fact, British troops planned to raid the storehouse. But the soon-to-be rebels caught wind of *that* fact and made a plan for their Minutemen:

```
gather on the high ground in Concord, MA
defend the munitions
```

Their plan anticipated that the British soldiers in Boston would take a slow route across the Charles River via a narrow strip of land known as Boston Neck. This plan would not work if British troops crossed the river more quickly by boat because they'd arrive in Concord before sufficient American soldiers would be in place. The name "Minutemen" was partly chosen aspirationally. The colonists needed a different plan in that case:

```
gather quickly on the high ground in Concord, MA
defend the munitions
```

Since they didn't know which of these two possible worlds would become reality, they needed a plan that would work either way:

```
if the British troops are crossing the Charles via Boston Neck:
    gather on the high ground in Concord, MA
if the British troops are crossing the Charles via boat:
    gather quickly on the high ground in Concord, MA
defend the munitions
```

The "if" is an example of a *meta-instruction*. This word signals the need to check the condition that immediately follows the *if* and, if that condition holds, carry out the commands indented below it. If the condition does not hold, skip those indented instructions and move on to the next set of commands. In this case, that's another *if* statement with its own set of indented commands to execute in case the British are coming sooner.

These instructions leave out the question of how the Massachusetts-ians . . . the Massachusetts-ers . . . the . . . okay, the people defending Concord would know which route their attackers were taking. That's a story that involves a little spying, a pair of lanterns, and another historically significant *if* statement: "one if by land, two if by sea."

In the Concord example, the *if* statements came in a pair—essentially describing what to do if the condition holds and what to do if that same condition does *not* hold. But consider the command (tautological exhortation, really):

Always be yourself!

The musician Jayy Von Monroe thought it would benefit from some special-case conditional tailoring:

Always be yourself! Unless you can be Batman, then always be Batman.

We can write this advice as:

```
if you can be Batman: always be Batman
if you can't be Batman: always be yourself
```

This pattern of what to do in the two separate hypothetical worlds is common enough that it gets its own shortcut in normal language and when programming: the *if-then-else* block.

```
if you can be Batman: always be Batman
else: always be yourself
```

We encounter *if-then-else* statements very frequently, although some events seem to cluster them. For example, weddings are a big choice point and are chock-full of *if-then-else* situations:

- If you want the beef, then check here, else you are getting chicken.
- If you are part of the bride's family, then please sit on the left, else right.
- If you know any reason that would prevent this couple from being legally married, then speak now, else forever hold your peace.
- If you catch the bouquet, then you are next in line to be married, else there will be six more weeks of winter

Quite often, there are more than just two possible worlds to consider. If we include an *if* statement *inside* the *else* command we can chain the conditions together and match up the right instructions with the right condition. For example, students of English as a second language are taught to produce the past tense of regular verbs using a set of rules like this (illustrated with some recent additions to our lexicon):

Rule 1: For most verbs, add -ed.

Examples: deplatform → deplatformed, free solo → free soloed, microtarget → microtargeted.

Rule 2: If a verb ends in e, just add -d.

> Examples: self-quarantine → self-quarantined, body-shame → body-shamed.

Rule 3: If a verb ends in a consonant and -y, take off the y and add -ied.

> Example: air fry → air fried.

Rule 4: If a verb ends in a single vowel (other than e) and a single consonant (other than x), double the consonant and add -ed.

> Example: Instagram → Instagrammed.

Consider what would happen if we translated these rules line by line into code:

```
add -ed
if the verb ends in e: add -d
if the verb ends in a consonant and -y: take off the y and add -ied
if the verb ends in a single vowel (other than e)
     followed by a single consonant (other than x):
   double the consonant and add -ed
```

When teaching children, "add -ed" comes first, but in code it can really mess things up because it would be followed unconditionally, essentially preventing the other rules from applying. Because that rule is our catch-all, it needs to be part of an *else* statement. But where? If we stick it onto any one of the *if* statements, then it'll catch the case where that specific condition doesn't hold. If we stick it onto *all* the *if* statements, then it applies multiple times for a verb, which would have never workedededed. We want it to apply only if all the other conditions fall through. We could handle that by putting a condition onto the "add -ed" rule that is the *conjunction* of the *negations* of all the other conditions:

```
if the verb doesn't end in e
   AND the verb doesn't end in a consonant and -y
   AND the verb ends in a single vowel (other than e)
     followed by a single consonant (other than x):
   add -ed
```

So, that works, but . . . ick. A cleaner approach is to use an *else-if* combination:

```
if the verb ends in e:
   add -d
```

```
else if the verb ends in a consonant and -y:
   take off the y and add -ied
else if the verb ends in a single vowel (other than e)
       followed by a single consonant (other than x):
   double the consonant and add -ed
else:
   add -ed
```

Now the rules work together as a block. Rule 2 is checked first. Only if rule 2 doesn't apply is rule 3 checked. If neither rule 2 nor rule 3 applies, rule 4 is checked. And if none of these three rules applies, *then* we finally apply the catch-all rule 1 and add -ed. The *else-if* idea lets us capture this notion of a domino chain of condition checking.

Later in the chapter we'll look at more deeply nested *if* statements that can be visualized as a branching tree of possibilities. But first let's look at an application of conditionals in the context of storytelling. Read on—if you dare!

CHOOSE YOUR OWN CONDITIONALS

Conditionals play a big role in choose-your-own-adventure books. In the typical book, each page of text follows the previous unconditionally. In a choose-your-own-adventure book, the reader is asked at various points along the story to select among different possibilities for how the story unfolds. For example:

It's a dark and stormy night and Sidney needs her sleep for the big audition tomorrow. She couldn't have been asleep for more than fifteen minutes when she hears a scratching sound coming from outside her bedroom window. She probably wouldn't have even noticed it over the whistling of the wind and the drumming of the rain, but there was something sinister about the sound. It was *insistent*, somehow. It was probably just tree branches bumping against the glass because of the rain, but she pauses a moment to consider what to do.

If she decides to go and investigate, go to page 7.

If she rolls over and goes back to sleep, go to page 9.

The italics mark the *if* meta-instructions as something apart from the story, instructions to the reader about how to proceed depending on which choice is selected. Story lines split, merge, and can result in different endings.

With the help of computers, the slightly awkward "go to page" instructions that move the story along can be replaced with a more seamless user experience, such as hyperlinks:

She considers <u>getting up to investigate</u> or simply rolling over and <u>going back to sleep</u>.

Beneath these clickable phrases, an implicit "if" lurks. *If* the player clicks on "getting up to investigate," the computer will show the part of the story corresponding to investigating. But *if* she clicks on "going back to sleep," the computer will proceed to convey what happens in that eventuality. (I guess there's a third choice where the player doesn't click on anything, but that situation is already familiar in the original choose-your-own-adventure books, too, aka—"choose-to-have-no-adventure-and -just-stare-at-the-page.")

A free, publicly available piece of software called Twine exists to support your choose-your-own-adventure authoring needs. I assume that the name comes from the idea that writing this kind of interactive fiction (sometimes, appropriately enough, called IF) feels a lot like becoming one of those obsessive investigators who create a "murder board," hanging photos of all the possibly relevant clues on the wall and then connecting them together in a tangle of string.

To better understand conditionals and how they can be used, let's take a closer look at Twine. The central object in Twine is the "passage," corresponding to a scene in a video or a chunk of story in text. A passage can include phrases that, when clicked on, move the story to another passage. The author can mark the linked text and the passage that should result by enclosing them in double brackets with a bar separating the text from the passage identifier. For the stormy night story, that might look like:

```
She considers [[getting up to investigate|page7]] or simply
rolling over and [[going back to sleep|page9]].
```

Let's flesh things out a bit more. For the stormy night story, going back to sleep moves the plot to the next morning (passage "page9"), when the rain has stopped and the sun is up. Sidney discovers that her bike is missing from the garage. She considers calling her dad, her mom, her step-dad, or the police. In all these cases, the calls set in motion a story-line in which she investigates the missing bike. However, investigating (passage "page7") reveals that the scratching at the window resulted from her friend Elana attaching a note to the window for Sidney to find. The note explains that Elana is involved in some kind of trouble and may need some help. Sidney looks around, sees no sign of Elana, and has no choice but to go back to sleep.

The story then picks up the next morning (passage "page7") again. When she finds her bike missing, she again has the option of calling her dad, her mom, her stepdad, or the police. But another option is to call Elana. Following this option results in a storyline in which she starts a search for her friend Elana. How, in Twine, can we make the option of calling Elana available, but *only if* Sidney had investigated the scratching sounds?

Twine includes a smaller unit of text than a passage: a "hook." The inclusion of a given hook in a passage can be made conditional using a special keyword called (perhaps you guessed it): "if." Here's an example:

Scat-a-long Fred walked into the arena and surveyed the crowd. (if: visits is 1)[It was his first rodeo and he was so excited to be there!] (if: visits > 1)[It was not his first rodeo.]

There are two hooks in this example, each inside a pair of brackets. The parenthetical expression before each one is its condition. If the condition is fulfilled, then the story includes the sentence in brackets.

The word "visits" here is used by Twine to keep track of the number of times the passage has been shown to the reader. It can be used with the *if* statement to control what text is displayed. As a result, the first time this passage is encountered, it reads:

Scat-a-long Fred walked into the arena and surveyed the crowd. It was his first rodeo and he was so excited to be there!

However, if this same passage comes up again in the story, it reads:

Scat-a-long Fred walked into the arena and surveyed the crowd. It was not his first rodeo.

The *if* statements tailor the text to the situation by including or skipping over hooks as appropriate. You can test whether the number of visits is equal to a particular number, not equal to a particular number, bigger or smaller than a particular number, or even a repeating pattern, such as being odd or even. (Those are really useful for helping to keep repeated text from becoming too repetitive.) You can even include coin flips in the story:

```
It was quiet. In the distance, (if: (random: 0, 1) is an even
number)[a dog barked](else:)[crickets chirped].
```

Here, the condition `else:` is just shorthand for the negation of what appeared before it. The code (`random: 0,1`) stands for either zero or one with equal probability. The test `is an even number` just checks whether the number before it is even. Randomly picking between zero and one results in an even number about half the time because one is not even but zero is. So the passage could end up being:

It was quiet. In the distance, a dog barked.

Or else:

It was quiet. In the distance, crickets chirped.

Ominous!

We can make a link that appears only if a previous passage was already part of the story by employing a Twine function to ask a question about the history of the current reading of it:

```
(if:(history: where its name contains "page7")'s length ≥ 1)
[[call Elana]].
```

It says to check for names of passages in the history (`history:`) and compile a list of all the times the "investigation" passage was visited (page7). If the length of that list is at least 1, that's because the story did in fact make passage through the passage in question. In that case, a `call Elana` link is included in the story.

In 2017, streaming Goliath (or streaming David, if you are Blockbuster Video) Netflix started offering a modern update on this kind of participatory story. Its 2019 release, *Bandersnatch,* an interactive episode of the series *Black Mirror,* won an Emmy for its forward-looking return to 1980s-era storytelling. In the show, the lead character is converting a classic choose-your-own-adventure book into a computerized interactive experience. Fantastically meta! As best I can tell, the authoring technology used by Netflix is built on top of Twine. For *Bandersnatch*, the millions of possible ways the viewer can move through the story made it hard for the filmmakers to keep everything under control. The software engineers at Netflix offered to help by extending Twine into a tool they called—and I tip my hat to them for it—"Branch Manager." (Huh. I just noticed that "Bandersnatch" is an anagram of "set and branch." Is that another programming reference? Maybe I should start a murder board to keep track of clues . . .)

Interactive fiction is a great opportunity to play with conditionals in a way that you might find fun and useful. Another is home automation, which we'll explore next.

THE FOURTH R

The ancient Egyptians built wooden door locks four thousand years ago consisting of a mechanism mounted in a box on the door and a bolt that prevented the door from opening.

The bolt had a set of holes placed precisely to match a set of drop-down pins in the mechanism. When the bolt was slid into position across the door, the pins lined up with the holes, allowing them to fall into place. The pins prevented the bolt from sliding out, securing the door. The key was about the size and shape of a toothbrush. It had teeth extending upward at precisely the same positions as the pins and holes in the lock. Slipping the right key into its slot and lifting up pushed the pins back out of the bolt and into the mechanism, providing an opportunity to slide the bolt out and free the door.

Because of this design, you could lock your door and leave it with a simple program:

```
if someone comes who has the key: let them in
else: stay locked
```

This, ladies and gentleman, is the original *if-then* machine.

Of course, these locks were not particularly programmable; even changing the "combination" required rebuilding the mechanism, bolt, and key. But the Egyptians were on to something. A similar design is used in modern pin-tumbler locks.

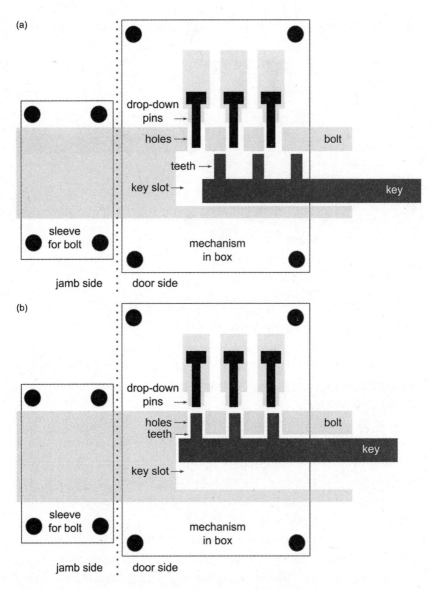

Figure 4.1 The ancient Egyptians built keyed locks for their doors.

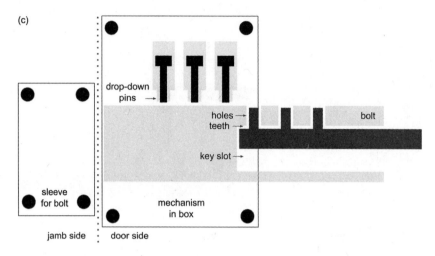

(c)

drop-down pins →

holes →
teeth →

bolt

key slot →

sleeve for bolt

mechanism in box

jamb side : door side

Figure 4.1 (continued)

Back in 2005 or so, I ran a seminar at Rutgers to explore the idea of making computer programming more universally accessible. We came to the conclusion that home automation applications, like the door-lock program above, might be a sweet spot—a use case that people would find compelling and conceptually simple and might motivate them to create programs themselves.

We built a collection of programmable devices, including lamps, buttons, timers, alarm clocks, motion sensors, coffee makers, fans, and a toaster oven. We were kind of new to hardware design and managed to short out the entire computer science building's electrical system at one point. (Okay, at two points, but it was at two different universities, so no one else knew it happened twice. Please don't tell!)

Fortunately, around the same time, computer-based controllers were getting smaller and cheaper. Companies like Hue and Wemo took advantage of these trends and began designing and selling "smart" light bulbs, outlets, and more. The devices could talk to your phone and the rest of the internet by way of your home WiFi, making it relatively simple to connect a bunch of them together. Many people refer to collections of connected devices as the "internet of things," or IoT. Such devices are much more safe and robust than the prototypes we built in the lab. But they are still a little below what you'd expect from mass-market appliances, so some people prefer to call it the internet of s**t.

The vision, though, is one I find quite compelling. Simple collections of *if-then* rules called *trigger-action programs* can coordinate the behavior of a wide array of devices and internet services. Today, it's popular to use Amazon's Alexa or Google's Home interface. But early on, the most comprehensive system available was a site called IFTTT ("if this then that," pronounced to rhyme with Lyft or gift or maybe miffed, if you are feeling frustrated). For a while, I followed the rule that I'd buy any device that was compatible with IFTTT. You know, for research. I officially ended the rule the day that BMW added its in-car control system to the site. I just couldn't come up with a good answer to "Sorry, could you repeat that? It sounded like you said you took out a loan to buy a new $100k car because a home automation website told you to." (By the way, IFTTT is still very much in existence.)

To give you a sense of the scope of possibilities, here are some trigger-action rules that I've created in IFTTT over the years, along with my personal ratings. (Four stars means it's a hit, zero stars means it's just s**t.)

- **if** Mobile Location reports that you exited the parking lot at work: Wink Nimbus sets its dial label to "Left work"

This rule was one of a set that I created that used the GPS on my phone to update a mini-programmable dashboard (a product called the Wink Nimbus) in our living room so it would show my current location. It acted like the Weasley Clock from *Harry Potter*, giving my family up-to-the-minute information about my whereabouts at a glance. Ultimately, the Wink folks started charging for their service, and I let it lapse. Four stars; would do it again.

- **if** Alexa reports hearing "sleep": Wemo turns dining room outlet off

The Alexa home speaker does a lot of home automation stuff directly, but it also has its own IFTTT link. I set it up so that if you say "Alexa, trigger 'sleep,'" Alexa sends a message to IFTTT. IFTTT reacts to the message by shutting off the WiFi-enabled outlet I had plugged the Alexa into, effectively putting the device to sleep. There was a *Star Trek* episode in which Picard defeated the Borg pretty much exactly like that. Four stars, as it's my last defense to preventing Alexa from assimilating all the devices in my home.

- **if** Whistle reports the dog's activity level hit two hours: Hue blinks the downstairs lights blue

I really thought this would be a good idea. I bought an animal-friendly activity monitor, called Whistle, and put it on the family dog. If she was active for two hours during the day, the lights would flash as a kind of celebration, or maybe an incentive. At least, that's what it was supposed to do. The device revealed that the dog is only slightly more active than a dining room chair, spending twenty-three and a half hours each day sleeping. The rule never triggered, and eventually my wife made me remove the device from the dog because she was worried it was too much of a strain for the dog to carry it around. Two stars, but that's actually for the dog because she's no quick brown fox.

- **if** Misfit watch reports button pressed:
  <u>Sengled turns off</u> living room lights

I was super-excited about this one. I bought the Misfit watch because the name made it sound like an off-brand Fitbit, which it kind of is. It's the cheapest home automation device I've ever bought and, I think, one of the coolest. In addition to telling the time and tracking my steps, it has buttons that transmit a signal to IFTTT. The rule above was something I created to show off for guests coming over for movie night. Once everyone had settled in, I'd say, "Let me turn off the lights," and I would push the button on the watch. Very superspy. Except the connection to IFTTT was through my phone, so if I didn't have my phone, it wouldn't work. And even if I had the phone with me, the corresponding app needed to be active. So I'd push the button on the watch, nothing would happen, and I'd say "Oops, wait a sec," and hunt for my phone, try to find the app, restart it, push the button again, realize the app had stopped responding, . . . At that point, someone else would "helpfully" stand up and turn off the lights. Two stars, for making me look more like Johnny English than James Bond.

- **if** Mydlink reports motion in the basement:
  <u>Hue turns on</u> basement light

When I bring laundry to the basement to wash it, I'm carrying the basket and can't reach the light switch. I thought I might be able to use a motion sensor to automate turning on the light. Unfortunately, there's a bit of a delay as my motion sensor transmits information about my arrival to IFTTT's servers three thousand miles away, my rule gets triggered, IFTTT checks in with Hue, and the information wends its way back to my house to turn on the lights. At first I was willing to wait patiently, standing in the dark holding my family's dirty clothes for fifteen seconds, waiting for the light bulb to go on. But the day it took two and a half minutes finally broke me. Two stars, for making me accommodate the system instead of the other way around.

- **if** Date & Time reports it is 15 minutes past the hour:
  <u>Wemo light switch turns off</u> closet light

When my wife looks at her clothes in the closet, she often leaves the light on, expecting to be back shortly, only to get busy doing other things. It has become my job to be on the lookout for this situation and turn off the light. That seems like a reasonable task to automate, but unfortunately, our motion sensor reports motion but doesn't report lack of motion, so it can't be used as a trigger. My workaround was to send an "off" signal periodically to the switch. But that meant it would sometimes take an hour to turn off the light, while at other times the light would go off while my wife was actually looking in the closet. One star—very frustrating not to be able to get the desired behavior.

- **if** GE Appliances Dishwasher reports a leak:
  <u>Gmail s\ends me an email saying</u> "check the dishwasher!"

IFTTT says I started the rule running in August 2020. It turned out that IFTTT and GE silently discontinued the service on a day about one month after that and one month before the dishwasher got clogged and flooded the kitchen. One star, thanks for nothing.

- **if** Wink Egg Minder reports that the egg count dropped below 2:
  <u>SMS sends a text to me saying</u> "time to buy some eggs!"

I didn't expect this rule, texting me to buy eggs when we were running low, to ever be all that useful. But in actuality, it never even worked. Zero stars, and I felt silly for putting eggs in an internet device inside my fridge.

Even if you don't have home automation devices, IFTTT can be really helpful for purely computational tasks. I have it alert me by email about newly published comics and tweets from specific people so I don't have to spend energy or attention monitoring for these things myself. IFTTT is almost intimidatingly welcoming, with a rule-creation interface that uses simple colorful pictures as well as text in roughly 100 point sans serif font that screams "SEE? I'M NOT SO SCARY! TRY ME! RIGHT NOW!" It really comes across as the kind of thing kids ought to be taught in grade school, along with reading, writing, and arithmetic and the fact that only one of those words actually starts with "r."

One of the amazing things about trigger-action programming is that you can do so much with such a simple rule format: *if* <trigger> *then* <action>. Nevertheless, you start to notice limitations once you get familiar with the structure. For example, my friend Justin (remember him from the Wordle example in chapter 3?) has a more elaborate home automation setup than I do that blends, with varying degrees of success, IFTTT, Google Home, and management hardware from longtime industry leaders Homeseer and Leviton. Each of these four components has its own trigger-action programming interface. Among many others, he wrote a set of rules to handle what to do when the basement door opens. If it's during the day, there's nothing to do. But at night, it turns on the lights. Which lights? Well, if someone is coming home after being away, the basement lights should come on. But if someone is at home and is heading out, the driveway lights should come on. Here's how he expresses this idea:

- **if** basement door reports being opened
     AND it's evening
     AND the alarm system is set to "away":
  <u>turn on</u> Basement lights <u>for</u> 30 minutes
- **if** basement door reports being opened
     AND it's evening
     AND the alarm system is set to "at home"
     AND driveway lights are off:
  <u>turn on</u> Driveway lights <u>for</u> 5 minutes

The important difference between Justin's rules and what I've done in IFTTT is the inclusion of AND as a connector between conditions. Knowing the basement door was opened is not enough information to decide what action to take, so he has the system check for a bit more context before acting: Is it night? Is the alarm system on? Are the lights already on?

The AND keyword is a little piece of logical notation that lets us express conditions more precisely. There are many logical operators like AND, but the big three are AND (also known as a conjunction), OR (also known as a disjunction), and NOT (also known as a negation). These are the main logical operators because they are very broadly useful. Even more fundamentally, it is known that any logical operation you could ever want to express can always be written down using combinations of these three operators. They are kind of the Swiss Army knife of conditionals.

For example, in the context of trigger-action programming, a disjunction could be used to say something like:

- **if** the doorbell camera recognizes Phyllis
     OR the doorbell camera recognizes Howard:
  <u>unlock</u> the front door

to unlock the door if either of my parents arrives. Most trigger-action programming systems don't allow for the explicit use of OR because you can get the same effect by splitting the rule into two:

- **if** the doorbell camera recognizes Phyllis: <u>unlock</u> the front door
- **if** the doorbell camera recognizes Howard: <u>unlock</u> the front door

It's a little more cumbersome but conceptually clean.

The use of negation is a somewhat messy in trigger-action programming because it's not quite clear how to interpret it. If I say:

- **if** the doorbell camera does not recognize Phyllis:
     <u>lock</u> the front door

does that mean the door should be relocked every second that Phyllis isn't standing in front of the door? Or only if it sees someone but that someone isn't Phyllis? To avoid this complication, specific triggers often have their own negations built into them ("temperature rises above"/"temperature drops below," "rain detected"/"rain no longer detected").

That means we can simulate disjunctions and negations using standard trigger-action rules. Conjunctions, however, seem pretty important to include explicitly in a trigger-action programming system. There would be no way for Justin to express the behavior he wants for his basement lights without them. When I first got interested in trigger-action programming, IFTTT didn't include conjunctions, although you can do fancier stuff now with their premium Pro and Pro+ packages.

Back in 2010 or so, my collaborator Blase Ur, who has a business casual name when said quickly, and I wondered why the IFTTT people left out conjunctions, in light of their importance. Our impression was that the designers thought conjunctions would be intimidating to users and their system was intended to be reassuring. We thought IFTTT was underestimating people, so we teamed up with a couple of undergrads and did a research study. We created an interface for authoring trigger-action rules and recruited people—nonprogrammers—to try it out. We found that over 80 percent of users we tested had no trouble constructing appropriate conjunctive rules like "If it is 10:00 p.m. and my bedroom door is closed and the lights are off, turn the television off," leading us to conclude that conjunctions pose no particular challenge for typical users. IFTTT shouldn't have been so conservative!

Alas, we spoke too soon. Later research by Maya Cakmak and Justin Huang pointed out that people have no trouble *writing* conjunction-based rules but were at a loss for how to *read* them. That is, people weren't sure what their rules would actually do.

The problem comes down to two things. One, like the key to a good joke, is the issue of *timing*. Triggers are events that happen at a particular moment in time. Ding! Your coffee is now finished brewing. Pop! The temperature in your room just went up to 75 degrees Fahrenheit. The conjunction of two events is itself, for all intents and purposes, a non-event. Specifically, there's no way that the millisecond the coffee finished brewing would be the exact moment that the temperature went from 74 to 75 degrees. To handle this issue, trigger-action programs combine an event with one or more "state conditions" that hold over longer intervals of time: "the coffee is currently ready" or "the temperature is currently 75 or above." If we write a conjunction for coffee being ready and the temperature being 75, we have to choose which will be the event and which the state condition. The fact that it matters is the second part of the problem with people reading these conjunctions.

Let's look at two different rules that we might give to a Jetsons-level automated house:

Rule A:

```
if coffee finishes brewing (event)
     AND temperature is 75 or above (state):
   announce "Coffee is ready. Don't forget to add ice!"
```

Rule B:

```
if temperature just hit 75 degrees (event)
   AND coffee is ready (state):
announce "It's getting too warm for coffee! Turn on AC?"
```

If either of the two rules (rule A or B) fires, it means that the coffee is ready and the temperature is 75 degrees or above. So you might think that if the coffee is ready and the room is warm enough, both rules will fire. But you'd be mistaken. That's because of the difference between an event and a state. I admit it gets a little hairy, but I find that it helps to think about two scenarios that can lead to both of these conditions being true.

Let's say the *event* of the coffee finishing brewing occurs at 7:05 a.m., but the temperature doesn't reach the *state* of being 75 degrees until 8:03 a.m. In this case, rule A won't fire, even though it's warm and there's coffee. The brewing-finishing event was a trigger that caused the system to check the temperature state at 7:05 a.m. After that time, it doesn't matter what the state of the temperature is, because the brewing event is over and done. So if rule A fires, you can be confident that the coffee brewed and the room hit 75 degrees. But if rule A didn't fire, it might still be the case that the coffee brewed and the room is warm; it's just that the event occurred after the state was achieved.

In contrast, rule B will fire in the example scenario. After all, the *event* of the temperature reaching 75 degrees happens at 8:03 a.m., when the *state* of coffee being ready is already true. Rule B will fire at 8:03 a.m.

In a second scenario, the temperature rises to 75 degrees at 6:45 a.m. That event triggers rule B to check whether the coffee is ready. It's not, so rule B goes back to sleep. Twenty minutes later, at 7:05 a.m., the coffee finishes brewing. This event causes rule A to check the temperature, which is in the right range. So rule A fires this time even though rule B didn't.

If this seems a little confusing to you, you are not alone. When we asked people in a follow-up study to distinguish between these rules, many of them (typically one-third or more) weren't able to see the difference. No matter how carefully we worded the text, they thought rules A and B would both fire any time it became true that the coffee was ready and the temperature was high enough. That's because they weren't seeing that one was an event and the other was a state. Although this distinction matters to the computer, it's lost on many people.

I found these results humbling. First, I felt bad for thinking the designers of IFTTT were short-sighted when they opted to leave conjunctions out of their system. I didn't realize that there are subtleties here that are not so simple to overcome.

Second, the result reinforces the message that creating interfaces that let people tell machines what to do is not just a matter of "if you build it, they will come" (another famous conditional). It's important to find ways to meet people partway. Maybe machine learning can help . . .

PAIR CONDITIONING

Back in 2005, I became convinced that the US Constitution has a logic bug. Here's the eligibility rule for being a US senator, as enshrined in America's founding document:

No Person shall be a Senator who shall not have attained to the Age of thirty Years, and been nine Years a Citizen of the United States, and who shall not, when elected, be an Inhabitant of that State for which he shall be chosen.

There's a lot of logic in there—specifically NOTs and ANDs. It's generally interpreted to say that, to be a senator, you have to meet three conditions: you must be at least thirty-one years old, you must have been a US citizen for at least nine years, and you must be a resident of the state you are being elected to serve. But it's a little more confusing because it's phrased in the negative, providing the rules for *not* being eligible for a seat in the senate.

But worse than the difficult-to-follow phrasing, I think the intent of the rule and what it actually says diverge. As a logical constitutional literalist, I have to conclude that the rule actually says that you only need to satisfy *any* of the three conditions, not all of them. Here's why.

Let's consider a simplified example with just two conditions. You are working as a ticket-taker in a movie theater and your manager tells you, "Don't admit anyone who is under eighteen and has no ticket." Madeleine asks to go into the movie. She tells you she is sixteen years old and doesn't have a ticket. She's definitely someone who is under eighteen and has no ticket, so you gently tell her that she won't be allowed in. Next, Jim, age fifty-six, walks up and presents his ticket. You should stop him only if he's under eighteen and has no ticket, so Jim can definitely enter. You scan his ticket and wave him in. All good so far.

Next, seventeen-year-old Joni flashes you her ticket. You ask yourself, "Is Joni someone who is under eighteen and has no ticket? Well, no. She's under eighteen, but she has a ticket. The rule says that means it's okay." You let her in. Claire asks to go in next, but she doesn't have a ticket. On the other hand, she's fifty-one years old. Is she someone who is under eighteen and has no ticket? No, she's not. She's *over* eighteen and has no ticket. Again, you wave her through.

So, following the rule that you should block people who are under eighteen *and* have no ticket, you can let in anyone who only has one of these shortcomings. AND means both. (Epilogue: You patiently explain your reasoning to your manager later, using well-established axioms and algebraic notation, and are promptly fired. Alas, life is not always logical.)

With this in mind, let's look at the constitution again. The structure matches the ticket-taking example:

No Person shall be a Senator who (*Don't admit anyone who*) shall not have attained to the Age of thirty Years, and been nine Years a Citizen of the United States (*is under 18*), and who shall not, when elected, be an Inhabitant of that State for which he shall be chosen (*and has no ticket*).

Following the same reasoning, the wording is that *either* being thirty and a nine-year citizen of the US *or* an inhabitant of the state is sufficient to avoid being blocked from becoming a senator. So, following the letter of the law, a teenager from Alabama can be a senator from Alabama, and a sixty-year old lifelong Texan can be a senator from Maine.

I posted this analysis on my web page back in 2005 and a former coworker, Pat Hawley, told me he found an error in my claim that there's an error. He pointed out that we should probably read this passage as an example of *ellipsis*. Specifically, consider the following (logically correct but awkwardly worded) passage:

No Person shall be a Senator who shall not have attained to the Age of thirty Years, and been nine Years a Citizen of the United States, and **No Person shall be a Senator** who shall not, when elected, be an Inhabitant of that State for which he shall be chosen.

Pat argued that this passage is the intended one, and the duplicated words were removed because they are implied by the context and are therefore redundant. Maybe so. I'll leave it up to you to decide. My point, though, and it's possible I made it more confusingly than is necessary, is that reading and writing logic can get quite tricky. Is there anything we can do to make it easier?

In chapter 3, we worked through an example of how you can get the computer to write a sequence of commands on your behalf. To adopt this kind of machine learning approach, you need three components:

- A representational space: What are the specific slots the computer will fill in for you?
- A loss function: How will the computer judge if a specific way of filling things in is good?
- An optimizer: How will the computer wade through the many, many, many possible ways of filling things in to find one with a good score?

In the context of conditionals, we'd like the computer to construct a block of nested *if* statements that associate the right *"thens"* and *"elses"* with the right *"ifs."*

For example, every morning while drinking your coffee you'd like to read a few stories from the daily newspaper. Many hundreds of stories are written each day, so skimming all the headlines to pick the few you want to read is a real drag. As in the recommendation example in chapter 2, you'd like to automate the filtering a bit so that the computer can take care of removing from the list headlines you simply don't need to see. Conditionals are a great fit for this problem. You might write:

```
if headline contains the word "dinosaur"
AND author is Lynn Arditi: show the headline
else: do not show the headline
```

because you are interested in dinosaurs and you are also interested in anything your journalist friend Lynn publishes. After a few weeks of getting no stories to read, you realize that you made a mistake. Even though you are interested in both of these kinds of articles, AND is the wrong logical connective to use. You are actually interested in seeing an article if it is about dinosaurs *or* if Lynn wrote it. What you had expressed instead is a desire to only see articles that Lynn has written about dinosaurs. Since she covers the health care beat in Rhode Island, the chances of her writing about dinosaurs are pretty slim ("COVID Cases among Stegosauruses in Pawtucket Spiked This Week after Having Tailed Off").

So you decide you want some help writing your filtering condition. The representational space will be nested *if* statements based on words in the headline or byline. The rule above will be written:

Rule A:

```
if headline contains the word "dinosaur":
  if author is Lynn Arditi:
    show the headline
  else:
    do not show the headline
else:
  do not show the headline
```

Rule A captures the AND version of the rule you mistakenly wrote. The machine will check the first *if* statement. If the headline contains "dinosaur," it will now execute the command indented below it. In this case, that command says we need to do another check: is Lynn the author? If so, the headline will be shown. If the author is not Lynn, however, the command after that first *else* will be executed, suppressing the headline. On the other hand, if the headline doesn't contain "dinosaur" at all, the final *else* is executed, and again, the headline is suppressed. The only way rule A shows the headline is if both conditions are true (a conjunction).

But that's not actually the behavior you wanted. Here's what you intended:

Rule B:

```
if headline contains the word "dinosaur":
  show the headline
else:
  if author is Lynn Arditi:
    show the headline
  else:
    do not show the headline
```

Rule B tests whether "dinosaur" is in the headline and, if it is, immediately shows the headline; that's enough to make the story relevant. If "dinosaur" isn't in there, rule B checks whether Lynn is the author. If so, again, we know right away that the headline should be shown. If neither condition holds, the headline is suppressed. Either condition being true is sufficient to show the headline (a disjunction).

In the context of machine learning, nested *if* statements like rule A and rule B are called *decision trees*. That's because each decision (does the headline contain "dinosaur"?) results in a branch, so the overall rule looks like an upside-down tree rooted at the first *if* statement. A decision tree could be a single *if* statement or it could involve hundreds and hundreds of branches working together.

With this representational space in place, you'll need to define a loss function so that the computer can distinguish good rules, like rule B, from undesirable ones, like rule A. The field of machine learning uses a clever trick to convey which rule is better based on how the rules *behave*, without having to solve the hard problem of scoring the rules themselves abstractly. You provide *explicit examples* along with how you want the rule to handle each of them. You might gather a collection of articles like the following:

Show these:

1. This Dinosaur Found in Chile Had a Battle Ax for a Tail (by Asher Elbein)
2. A Newly Discovered Dinosaur with Sharklike Teeth Was the *T. rex* of Its Day (by Katie Hunt)
3. Attorney General: Rhode Island's Health Care System Needs More Than a One-Deal Solution (by Lynn Arditi)
4. Former Washington Health Official Returns to R.I. to Lead Eleanor Slater Hospital (by Lynn Arditi)

Do not show these:

5. Acciona Wins $258m Hospital PPP Project in Chile (by David Rogers)
6. The 3,500-Year-Old Mummy of an Egyptian King Has Been "Digitally Unwrapped" for the First Time (by Katie Hunt)
7. Omicron Is Sidelining Even Health Care Workers as It Rips through Texas (by Ren Larson)

The computer can use these examples to score the two rules. Rule A would not show any of these seven articles, including the four that we wanted. So, in terms of accuracy, the rule gets three out of seven, or about 43 percent correct, for appropriately suppressing articles 5 through 7. In contrast, a rule that says to show any article with "health" in its headline would show articles 3 and 4 (correctly) and 7 (incorrectly), to obtain an accuracy of four out of seven, or about 57 percent. Finally, rule B correctly categorizes all seven articles, for an accuracy of 100 percent.

Sifting through all possible decision tree rules to find one with high accuracy is known to be a very hard computational problem. Nevertheless, many highly successful optimizers have been built and are available online. One thing that makes the problem especially tricky is that, all things being equal, we prefer small rules to big hairy ones. This idea is well known in scientific circles as "Occam's razor," which holds that "entities should not be multiplied beyond necessity." It is generally taken to mean, Don't make things more complicated than they need to be. In the context of learning decision trees, that means: Find a rule that is accurate but also has as few *if* statements as possible.

As a little experiment, I downloaded a decision tree optimizer and applied it to a bunch of titles of scientific papers. Half the papers came from a machine learning conference (which I find very interesting) and half came from a robotics conference (which tend to be less relevant to my work). I asked the optimizer to find the most accurate decision tree to distinguish the robotics and machine learning papers based on their titles. (It's similar to a game I play where I guess the news source from the headline my iPhone shows me: "Chris and Katherine Schwarzenegger Pratt Expecting Baby No. 2," "New York Says It Will prioritize Non-White People in Distributing Low Supply of COVID-19 Treatments," "39 Things under $25 To Help You Spruce Up Your Home Decor in 2022" → *People, Fox News, BuzzFeed*.) I also asked the optimizer to avoid nesting *if* statements any deeper than two levels so that the resulting rule would be readable. Here's what it came up with:

```
if title contains "of":
    if title contains "learning":
        guess it's a machine-learning article
    else:
        guess it's a robotics article
else:
    if title contains "robot":
        guess it's a robotics article
    else:
        guess it's a machine-learning article
```

Some of the rule seems spot on. It will guess that any article with "learning" but not "robot" is machine learning and any article with "robot" but not "learning" is robotics. (Researchers are a bit like Poké-mon: they like to say who they are right up front.) But what if it has neither? In such a case, the rule checks whether the word "of" is in the title. If so, it guesses robotics. Why? To be honest, I'm not sure. There's no question that the word "of" is informative in this context, but it's not at all clear whether there is a deep or lasting reason for it. That's one of the challenges of using automatic methods to tell machines what to do. The methods are almost certainly better than you are at finding patterns in data and using those patterns effectively. But there's no guarantee that the patterns are anything but accidental. If a learned rule for filtering headlines does something weird, it's probably not the end of the world. But if the rules are being used to, say, advise a judge as to whether or not a criminal is likely to commit a serious crime if released on bail, well, the consequences for society can be quite significant. We'll return to this concern in chapter 8.

For now, however, it's worth noting that conditionals are central to being able to tell machines what to do, and using examples to create conditionals greatly expands the set of options. Decision trees learned from data are being used to make medical diagnoses, predict movie profitability, assess housing prices, identify spam emails, and much more.

In the next chapter, we'll talk about variables and how they broaden our vocabulary for telling machines what to do. We'll also see the role they play in machine learning and how they fueled a dramatic renaissance in applying AI ideas to problems in the real world.

5

STORING IN VARIABLES
IT'S WHAT I STAND FOR

Variables stand in for objects, which is a jump in abstraction that can seem disorienting at first. But this concept is familiar to all of us. Whenever we use words (which happens often when we're not performing mime), we are really working with variables. A common noun like "dog" can be used to stand in for any actual specific dog in the world. (My family's dog was adopted over my objection, so I don't consider her my dog. That's convenient when I'm walking her and she pees on the "Keep your dog off our lawn" signs in my neighborhood.)

In the context of programming, words like these are known as *variables* because what they refer to can vary. But this feature can also generate bugs in programs and in real life. For example, the first talk I went to when I arrived at Brown as a professor was a visiting lecture by Peter Dayan, a brilliant contributor to my home field of reinforcement learning. He sprinkled references to the great work I was doing throughout the talk and I felt very welcomed in my new institution. Weirdly, though, some of the results he attributed to me weren't things I had gotten around to doing yet. Eventually, it dawned on me that "Michael" is a variable. And it was standing in for not me but Michael Frank, a Brown neuroscience professor and Dayan's host for the visit. Oops.

It gets worse. Years before, when I was working as a visiting professor at Princeton, I was delighted to get a reimbursement check mailed to me after my first week, only to discover that it was actually intended for someone else. The check was for Professor Michael Littman teaching about computational methods for decision-making in Princeton's School of Engineering and Applied Science in the department of Mechanical & Aerospace Engineering, *not* Professor Michael Littman teaching about computational methods for decision-making in Princeton's School of Engineering and Applied Science in the department of Computer Science. Yikes. Your parents tell you you are unique and special, but sometimes the universe strongly hints otherwise.

Despite the occasional ambiguity, variables come in handy in the kitchen. One of my life goals is to cook a Japanese hibachi meal. After all, hibachi involves two of my favorite activities: eating rice and juggling knives. I haven't done it yet, but there are some great instructions online that make it sound . . . well, not easy, but at least attainable. Hibachi recipes also make use of the three main ways variables are used in conveying tasks, so let's look at a few examples. The instructions start:

First, let's make some shrimp as an appetizer. Cook the shrimp with a half tablespoon of butter for two to three minutes on each side. Flatten with spatula while cooking. When ready, throw the shrimp at your family, ideally into their mouths.

These instructions are quite similar to the sort of thing we looked at in chapter 3. There is a sequence of steps that we are being asked to carry out. But I want to draw your attention to the phrase "the shrimp" here. It refers to whatever shrimp we bought, caught, or brought to the kitchen. The same sequence of commands applies to any of those shrimp. We don't need to reexplain what to do for each possible shrimp (which is good, since there are billions in the world and more being added all the time). Instead, the phrase "the shrimp" stands in for the shrimp we're cooking—the *initial input* of the food-prep process. That's a major power of variables: they let you apply the same task to a universe of possibilities by connecting a specific name ("the shrimp") with any of a large number of possible values (the shrimp you are cooking). The variable in this case is the phrase "the shrimp" because its value is allowed to *vary* from one application of the recipe to another. Continuing on:

Hibachi isn't hibachi without yum-yum sauce. Combine mayonnaise, ketchup, garlic powder, paprika, sugar, and tabasco sauce in a food processor and blend. Place mixture in a container and chill overnight so flavors can bloom. If you have any shrimp left over, you can dip them in the sauce. You can also pour the sauce over chicken.

The phrase "the sauce" is being used as the variable here. Like "the shrimp," it refers to a different real-world object each time you follow the recipe. The important difference between the yum-yum sauce example and the shrimp example is that the phrase "the sauce" refers to something that doesn't even exist when we start following the recipe. It is created as an *intermediate result* as part of the cooking process. Giving it a name makes it a "thing" and allows us to talk about its use and its interaction with other elements of the meal. Finally:

As a main course, we'll prepare some chicken. Start with boneless skinned chicken breasts. Cut the chicken into 1-inch pieces. Heat vegetable oil on the griddle, then add the chicken, laid out into a flat layer. Flip the chicken after two to three minutes. Add a little lemon juice at the end.

In this step, "the chicken" refers to a boneless chicken breast at one point and bite-sized chunks of chicken at another. The same name can refer to *transformations* of an item, or even entirely different items, over the course of the instructions. That's the third main way to use variables, having their values vary during the process itself. Put them all together and you get one tasty meal!

Let's take a brief look at how these three ways variables are used—initial inputs, intermediate results, and transformations—play out in traditional coding. Snow White, on whom the Seven Dwarfs have become quite dependent, needs a day to herself. The Dwarfs typically ask her to mediate all kinds of simple decisions, like taking a majority vote for whether or not the Dwarfs will go on a group hike. So she decides to write a simple program for Doc, Grumpy, and Dopey in case they need help while she is away. The program makes reference to the *initial input* information in three variables that represent each of their hike preferences yes-or-no: `doc wants to go`, `grumpy wants to go`, and `dopey wants to go`. If at least two of these three miners want to go, the hike will happen. If not, not. She can write the rule by referencing these three variables using the *if-else* idea from chapter 4:

```
if doc wants to go AND grumpy wants to go: say "the hike is on!"
else if doc wants to go AND dopey wants to go: say "the hike is on!"
else if grumpy wants to go AND dopey wants to go: say "the hike is on!"
else: say "hike canceled due to insufficient interest"
```

If any of the pairs wants to go, majority rules! If all of the *if*s fall through, that means either a single individual wants to go or none of them wants to go. So that code works.

Snow White realizes that this solution, while reasonable for a trio of Dwarfs, doesn't scale well to the full seven. She'd need a program with thirty-six lines, one for each of the thirty-five distinct groups of four Dwarfs that could constitute a majority, plus one for the *else* at the end. That's not great. Writing out the thirty-five combinations is tedious and error prone. ("Wait, why do I only have sixteen lines? Oh, darn, I always forget Bashful!") And if she were to have to do the same thing for the 101 Dalmation neighbors, she'd need 199,804,427,433,372,226,016,001,220, 057 lines. So much for her spa day.

She gets another idea. She knows she can use variables to represent *intermediate results* in a complex calculation. The most useful intermediate result for this problem is: how many Dwarfs want to go? If she knew that count, a simple *if* statement would suffice to determine whether more than half want to go and therefore whether the hike should happen. She defines three new variables: *doc count*, *dopey count*, and *grumpy count*. Each of these variables takes on the value one if the corresponding Dwarf wants to go and zero otherwise. Then, Snow White can create a variable *group count* that is the sum of these three variables, making the hiking decision based on this sum:

```
if doc wants to go: set doc count to 1
else: set doc count to 0
if dopey wants to go: set dopey count to 1
else: set dopey count to 0
if grumpy wants to go: set grumpy count to 1
else: set grumpy count to 0
set group count to dopey count + doc count + grumpy count
if group count > 1: say "the hike is on!"
else: say "hike canceled due to insufficient interest"
```

Much cleaner! And now even the Dalmation case can be handled in a few hundred lines of code. She decides to make one last improvement using the idea of *transformations*, in which a variable's value can actually change over the course of the calculation. Inspired by those little tally counter devices, she imagines marching the Dwarves past her one by one. The group count starts at zero and one is added to it, clicking it forward, for each individual who wants to go on the hike. With this approach, in the same number of lines of code that she used to handle a vote for the three Dwarfs, she can take care of all seven:

```
set group count to 0
if doc wants to go: add 1 to group count
if dopey wants to go: add 1 to group count
if grumpy wants to go: add 1 to group count
if sneezy wants to go: add 1 to group count
if happy wants to go: add 1 to group count
if sleepy wants to go: add 1 to group count
if group count > 3: say "group is going"
else: say "activity canceled due to insufficient interest"
```

Later in the chapter, we'll see other ways variables are used in more restricted programming settings.

YOU CAN'T ESCAPE SPAM

But first I want to focus on a somewhat challenging issue that arises in the context of variables. Variables stand in for things, but the names of variables are things themselves. So, in the Snow White example, "activity canceled due to insufficient interest" can be a message—a literal value. That's how it's used in the example. But in other contexts it could also be a *variable* that refers to an activity canceled due to insufficient interest—a hike, say. In everyday language, we might mark a phrase with how we want it interpreted if we think there might be ambiguity. I could ask you, "How long is Old Town Road?," and you could say the song, which topped the Billboard Hot 100 list for a record-breaking nineteen weeks in 2019, was only a minute and fifty-three seconds long. Or I could ask you, "How long is *the phrase* Old Town Road?," to refocus your attention on the three-word sequence (essentially the name of the variable) instead of the song it refers to (its value). The phrase "the phrase" serves as a marker to take what follows it literally, just as the quotation marks around "the phrase" tell you to take what's inside it literally.

Similarly, when we tell computers what to do, we have to mark whether we are using a name literally or as a variable referring to a value. Punctuation marks, such as quotation marks, brackets, backslashes, ampersands, and other annotations, are commonly used to distinguish these two cases. But it's easy to get it wrong, as you can sometimes see in the junk email you get.

Not long ago, I got an email with the subject line:

[[FIRSTNAME, OR "Good Morning,"]] are your WFH employees practicing good ergonomics?

I spotted the mistake right away. Can you? No, the sender did not misspell "WTF," although that's just what I said when the email arrived.

A local company called "Bisson Moving & Storage" wanted to send me a marketing message. And not *just* me. They have a huge mailing list of people like me that, I assume, had a moment of indiscretion and gave their email address to someone who should never have been trusted with it in the first place. You are officially the worst, Maine Human Resources Convention.

Anyway, Bisson registered with a popular email marketing site called Constant Contact. I'm guessing that's the name they chose because they didn't *want* to sound like they provide services for needy stalkers but stopped caring before they could come up with something better. Similarly, I thought it would be a good idea to change the names of the companies Bisson, Maine HR Convention, and Constant Contact to protect them from feeling as though I'm chiding them. But, after two years of their unsolicited email, I stopped caring before I could come up with something better.

Bisson wanted to tell thousands of people that "Bisson Moving & Storage" really cares about reducing workplace injuries. But they were worried that their message of deep concern would be undercut if it looked like they were sending it blindly to thousands of people. So they took advantage of a powerful personalization feature Constant Contact offers its customers. If you are a Constant Contact user, you can send, to everyone on your selected contact list, the email you write. However, before sending the message, Constant Contact scans it, looking for text enclosed in double brackets, like this:

[[this text is enclosed in double brackets]] .

The double brackets in a Constant Contact email signal a part of the message that should be treated differently from the rest. It's a kind of variable, in contrast to the literal text outside of the brackets. Instead of being sent along as written, these blocks act as requests to the Constant Contact system to replace what's in brackets with something else. That something else can be personal information you have on file about the message's specific recipient. By entering

How is your dog, [[CUSTOM.dog_name]]?

Constant Contact knows to substitute the name of that person's dog into the email to get How is your dog, Rover? if the recipient's dog's name is Rover—something we know from chapter 1 to be unlikely.

Constant Contact knows that not everyone on its user's contact list will have all the relevant data fields in its database, so it provides "fallback text" if the requested information isn't available for the recipient. Entering

[[FIRSTNAME OR "Good morning"]], how are you today?

as the subject line of the message asks the computers at Constant Contact to address the message to the recipient's first name, or, if you have no first name on file for that recipient, the mailer should just say "Good morning." So someone like me would get a message with the subject line

Michael, how are you today?

or

Good morning, how are you today?,

depending on whether they know my name.

Nice, right? Really makes you forget that you are being thought of as a low-probability "conversion opportunity" for some thirsty marketing team hundreds of miles away. At least that was Bisson's hope. Unfortunately, the company slipped up and showed us what's behind the curtain.

In this case, it looks like someone at Bisson realized that

[[FIRSTNAME OR "Good Morning"]] are your WFH employees practicing good ergonomics?

would lead to poorly punctuated subject lines like

Michael are your WFH employees practicing good ergonomics?

Standard usage (hey, *Chicago Manual of Style,* section 6.38!) says we need to include a comma, right? So the eagle-eyed Bisson employee added commas after the two possible greetings:

[[FIRSTNAME, OR "Good Morning,"]] are your WFH employees practicing good ergonomics?

Seems sensible enough. But the software at Constant Contact is kind of . . . particular. You can give it any text you want *outside* the double brackets. You can give it any text you want *inside* the quotation marks inside the double brackets. But any weird characters *inside* the double brackets but *outside* the quotation marks, well, that's just too confusing. Because of the comma after "FIRSTNAME" in Bisson's message, the software no longer has any clue how to interpret the entire message inside the double brackets. It "assumes" the message author didn't actually want to do a substitution (if they did, why would they possibly have put that unwelcome comma next to FIRSTNAME?) and instead intended it to be part of the literal message being emailed.

And, that's why, that morning, instead of getting the personal-looking

Michael, are your WFH employees practicing good ergonomics?,

I got the off-putting

[[FIRSTNAME, OR "Good Morning,"]] are your WFH employees practicing good ergonomics?

There are several lessons here. First, don't register for human resources conferences in Maine. The organizers seem friendly, but they are soulless monsters selling your private information to the highest bidder and also to all the other bidders and literally anyone who asks for it.

But more relevant to this chapter is the second lesson. When we tell machines what to do using variables, we have to distinguish between text to be taken literally and text to be taken as the name of a variable whose value should be substituted into the current context. In the examples I use throughout the book, I mark variables using italics and a fixed-width font and values sometimes by enclosing them in quotation marks or simply not italicizing them. Some other languages use backslashes or ampersands. For example, you can sometimes see `&lt;` on a web page if the variable that stands for the less than symbol (<) doesn't get properly interpreted.

I'm not sure why the spam message bothered me so much, but lest you think it soured me on variables, here's a similar but more lighthearted example. Mike Ward wrote a short piece riffing on this idea on McSweeney's humor site titled "E-mail Addresses It Would Be Really Annoying to Give Out over the Phone." Consider how you'd do it if you had to! And my apologies to the folks listening to the audiobook. Here goes:

- MikeUnderscore2004@yahoo.com
- MikeAtYahooDotCom@hotmail.com
- Mike_WardAllOneWord@yahoo.com
- AAAAAThatsSixAs@yahoo.com
- One1TheFirstJustTheNumberTheSecondSpelledOut@hotmail.com

See? Variables don't always take themselves too seriously.

BROADLY IN SCOPE

Let's apply these ideas to a concrete computational problem—checking whether a sixteen-digit credit card number is valid. Perhaps that's not something you personally need to tell a computer to do, but it's definitely something we'd rather do by computer than by hand. Here are the steps of the process:

1. Start with the sequence on the credit card.
2. Replace the digits in the sequence that are in odd-numbered positions (first, third, fifth, etc.) with the value you get by doubling them and then adding up the digits in the result. For example, a 7 becomes 14 and then 5.
3. Calculate the sum of all the digits in the sequence.
4. If the sum does not end in a zero, then the credit card number you started with is not valid.

This procedure, known as the Luhn algorithm, is designed to be fast and easy (at least for machines), and to flag single-digit errors (write down one digit wrong and the number will be invalid) and most digit swaps (if you flip around any two adjacent digits other than 09 or 90, the result will be invalid).

The Luhn algorithm also illustrates all three of the ways variables are used: initial input, intermediate result, and transformations. In step 1, the variable "the sequence" refers to the initial input—the same procedure can be carried out on any sixteen-digit number. Step 2 generates a few intermediate results, in the variables "the product" and "them," and uses the values to transform the sequence. Step 3 uses "the sequence," now transformed, to produce "the sum," which is also referred to in step 4. Without having these quantities named and called out, specifying this process would be completely impractical.

There's one other use of variables here that I'd like to bring to your attention. The phrase "the digits" appears twice in step 2, referring to two entirely different things. In the first context, they are the digits of the sequence. In the second context, they are the digits of the product (a doubling of an individual digit from the sequence). Even though it's the same name, it's really two entirely different variables. They are distinguished by their *scope*, which is essentially the context within the instructions where they are used. In this case, the inner scope is when we've doubled a digit and the product might consist of multiple digits (for example, because we doubled 7 and got 14). The outer scope is when we are considering all the digits in the credit card number itself.

Figuring out the relevant scope for a variable, or, even more important, figuring out how to express an idea so that the relevant scope is clear, is hard. After all, when authors write lots of sentences, it's easy for them to become mixed up. Wait, did I mean that the authors get mixed up? Or the sentences? Or are the authors being mistaken for sentences somehow? I'm not even sure what I meant at this point.

Even very smart people can get confused about scope. In February 2013, Stephen Colbert interviewed Justice Sonia Sotomayor on his show *The Colbert Report*. He asked her:

When you go to the Bronx and you see a young girl on a playground there, do you think, "That child has the same opportunities I had when I was a child"?

She paused, confused, and finally answered, "Well, a piece of me would want her to become a justice instead of a person on television."

There are a few things to notice here. One is that you can be a highly successful professional who has trained for years to use language as a precision instrument but still not know how to deliver a decent burn. Second is the first law of being interviewed: the host is a stand-in (a variable!) for the audience. If you attack the host you are attacking the people watching you. Don't do that.

But the deeper issue is the use of the personal pronoun "I" in Colbert's question. Personal pronouns stand in for people, and the person you should be substituting in when "I," "me," or "my" is used is the identity of the speaker. Simple, right? But because Colbert's question introduces a hypothetical scenario, there are two distinct speakers—two scopes—at play, Colbert as the interviewer and Sotomayor as the hypothetical visitor to the Bronx playground. The question makes it clear that the speaker of the question, the "I," is intended to be Sotomayor.

Actually, it's not that clear. The difference is one conversational "beat" after the word "think." If you remove the slight pause, the question becomes:

When you go to the Bronx and you see a young girl on a playground there, do you think that child has the same opportunities I had when I was a child?

Same exact words, but now the word "I" binds to, or references, Colbert instead of Sotomayor. When I listened to the interview, I was surprised by Sotomayor's misinterpretation. But should I ever be so lucky to be on Colbert's show (please???), I just hope I'm sharp enough to make such a subtle distinction on the fly.

Variables in computing, like words in language, can refer to different things at different times. (Maybe that's obvious from the name: if something can't change its value, calling it a variable seems like a violation of truth-in-advertising statutes.) But it's critical to disambiguate scope in both cases to avoid potentially catastrophic misinterpretations.

SHEETS HAPPEN

If there's one mechanism for conveying tasks to machines where variables are in the spotlight, it's spreadsheets. The whole interface is an enormous table of cells, each of which is a variable. The values of the variables appear inside the little boxes. You can just fill them in yourself (*initial inputs*), or their values can be computed as *intermediate results*. (The third main way to use variables, *transformation*, is not normally allowed in spreadsheets. If you ask a spreadsheet to do a transformation, you get a "circular reference" error.)

Because of the huge number of variables available in a spreadsheet, it doesn't make sense to give each one its own name, such as "FIRSTNAME" or "the chicken." Instead, spreadsheet designers took a cue from city planners, or maybe the game *Battleship,* and named each cell (variable) by the intersection of the street and avenue where it is located. The rows of the table are numbered sequentially starting at 1. The columns of the table are lettered starting from A and proceeding to Z, followed by AA through AZ, BA through BZ, and so on. The cell that's in the fifth row and fourth column gets the name "D5."

Naming the cells by position saves the trouble of thinking up unique names for each cell, but it also means that the names of nearby cells are predictable and that variables of a given type can be grouped together and operated on. What's two cells left and three cells down from T11? Simple. Just go back two letters in the alphabet from T and forward three numbers from 11: R14. To appreciate how useful it is to know the names of nearby cells, let's walk through a simple spreadsheet application. Stepping through this exercise with you will give you a sense of how working with these kinds of variables goes.

As a professor, one of my official duties is to assign letter grades to all the students, perhaps two hundred of them, in my classes. The TAs and I collect student scores during the semester in a spreadsheet where each row corresponds to a student. The first five columns give the student names and their grades for the different components of the course: homework assignments, class participation, midterm, final exam.

I'd like to use the spreadsheet from this semester to consider revamping how the course components contribute to the final grade.

There are particular students who, intuitively, I feel should have gotten A's but didn't, and a few who got A's but probably shouldn't have. This semester, I weighted homework assignments 40 percent, class participation 15 percent, midterm 15 percent, and final exam 30 percent. How many students would have gotten A's had I instead weighted the components evenly at 25 percent, 25 percent, 25 percent, and 25 percent?

I want the spreadsheet to compute final grades based on whatever weights I put there. That's a lot to think about all at once, so my best bet is to break down the calculation into simpler, bite-sized (byte-sized?) pieces. Hitting on a clean way of breaking up the problem is a bit (bit?) of an art form, but maybe we can think a little backward and a little forward and figure out a way to connect the two.

To sum up all the A's that would be assigned, I'll need a course score for each student. I give students an A if their course scores are at least 90 percent. The course score is a combination of the course component weights and the percentage the student got on each course component. I could calculate this intermediate result by hand for each student and put it into a new cell. But instead I'll put in *explicit instructions*—in this case, a formula—for computing the value for the first student, the one in the first row. I need to tell the spreadsheet to actually interpret these instructions so it doesn't just come back with Amelia-Bedelia-style literalness. The standard way to convey to a spreadsheet "please don't take me literally" is to put an equal sign in front of the formula. So here's what I type into cell F1, just to the right of the first student's scores:

$$= .25 * B1 + .25 * C1 + .25 * D1 + .25 * E1$$

It is taking the student's course component scores, weighting them by the proposed course component weights, and computing a total. Immediately, the first student's course score appears in cell F1. It computed our intermediate value!

I could type similar formulas for all two hundred students in the class, but the chances of taking more than an hour and making at least one mistake are pretty much 100 percent. Fortunately, the scheme of naming each cell with its address comes in handy here. If I paste the formula from the first row into the cell one row below it, the spreadsheet does an interesting thing. It interprets the reference to the scores and weights in *relative* terms instead of absolute terms! That is, it's not "the first student's scores" but "the scores for the student in the current row." The formula looks like this:

$$= .25 * B2 + .25 * C2 + .25 * D2 + .25 * E2.$$

Note the 2s in row 2 where there used to be 1s in row 1. It's as if the references to the variables (*B1, C1, D1, E1*) are themselves *variables* that can refer to other cells. Meta strikes again!

Pasting this formula into all the cells from F1 to F200 results in these same adjustments being made, and the calculations all take place automatically. *Voilà!* All the course score percentages have been computed. (The French taxi driver from chapter 1 would be proud.)

As a final step, we can paste a formula into any available cell, say G1:

= <u>countif</u>(*F1:F200*,">=90").

It says to make the value of the cell G1 equal to the number of cells in the range F1 to F200 that are at least (greater than ">" or equal to "=") 90. That is, it fills in the cell with the number of students who would get A's under the new weighting scheme. (Turns out to be one hundred eighty out of two hundred. I think I might be giving a bit too much weight to class participation . . .)

None of the operations in this example are all that complicated—addition, multiplication, and comparisons. Coming up with the right intermediate results to assign to the variables helps make all the calculations pretty simple. Is there some way we can get help choosing the right intermediate results? Indeed, there is. Like the decision trees discussed in chapter 4, neural networks are an approach to machine learning, but this time focused on defining the right variables (and their values). Let's take a look.

"AW, THEY'RE JUST LIKE US!"

A famous statistician once said, "Machine learning is just statistics done by people with better names for things." I thought that was a pretty clever way to put it, especially for a statistician. And he had a point. The fundamental problem of learning to make predictions, a key concept in machine learning, is at the heart of what statisticians have been doing since at least the 1920s. As a result, statisticians and computer scientists work on a lot of the same problems, but they often have different names for things. For example, a core technique in use in statistics since the 1950s called "logistic regression" is known in the machine learning community today as a "neural network." And "neural network," we can all agree, is a statistically significantly cooler name.

But neural networks have grown far beyond the logistic regression representation. Neural networks today allow computers to interpret audio signals as speech, to recognize the objects present in images, and to translate Japanese text into a readable English equivalent. Roboto-san, arigatō gozaimashita. (Thank you very much, Mr. Roboto.) They can draw images in a variety of styles just from natural language prompts. (A network called DALL•E 2 from OpenAI drew the Mars rover illustration in chapter 1 given the prompt "A robot rover on Mars, consulting a map because it got lost." Neat, right?) What they are doing is far cooler than what a logistic regression approach could accomplish given the same data. Computer scientists decided it's even cooler than what neural networks used to be able to do back in the early 1990s, so they chose an even cooler name to match. They are now known as "deep networks" trained by "deep learning." Oooh, deep!

Deep learning burst into the public consciousness in 2012. More specifically, the fact it had burst into the public consciousness burst into *my* consciousness that year when we were doing some home remodeling. A cabinet expert was working in my kitchen and talking casually to my wife. He said, "Did you know Google's computers have started watching cat videos?" I was shocked to realize he was talking about the work of my colleague, Andrew Ng, who led a team at Google Brain applying new deep learning algorithms to photographs. Or, as John Markoff of the *New York Times* put it:

Google scientists created one of the largest neural networks for machine learning by connecting 16,000 computer processors, which they turned loose on the Internet to learn on its own.

Presented with 10 million digital images found in YouTube videos, what did Google's brain do? What millions of humans do with YouTube: looked for cats.

There's something aggressively charming about the idea of an oversized electronic brain that is released onto the internet and becomes infatuated with Keyboard Cat. It's like imagining that the *Terminator* robot comes from our own dystopian future and falls in love with videos of puppies drifting off to sleep, ultimately deciding that destroying humanity would be a tragedy after all.

That said, this summary of the work is, unfortunately, misleading. The description of a bunch of computers "turned loose" is evocative, but wrong. The researchers collected thumbnail images from the internet and the machines used a computer network to share information, but the computers involved in the experiment were *not* on the open internet during the experiment.

And since they weren't interacting with the wider network, it doesn't make sense to say they were "looking for cats." More accurately, the computers were force fed cats, exposed over and over again to the same set of images until those images were seared into their simulated brains. It's less a story of computers surfing the web for pictures that delighted them and more the "Ludovico Technique" from *A Clockwork Orange*—you know, the scene where Malcolm McDowell's character's eyes are forced open as a series of images is flashed before him while his physiological reactions are manipulated to influence the associations he makes with the images. If that sounds like hyperbole, it's not. Over a period of three days, the program was repeatedly shown the same ten million pictures culled from YouTube videos. The program had to learn a way to translate each image into numerical patterns that could themselves be turned back into replicas of the original images. Or else.

The Google team discovered that one of the neurons—a variable in the deep network—specialized itself to detect when it was given an image of a cat. When exposed to a data set of images, each tagged by people as to whether it contained a cat or not, the most cat-aligned neuron was 74.6 percent accurate at flagging noncat pictures. By comparison, a dead neuron that never responded to any input would score 64.8 percent on this test because most of the images were not cats.

A snarky statistician might say, "Google needs 16,000 computers to get a 10 percent accuracy improvement over what a rock could do," but it was a start. Later generations of this idea are able to identify thousands of different visual categories at an accuracy approaching 99 percent. We can now delegate to machines the problem of recognizing important items based on their visual appearance. It's a long-standing goal that, prior to the development of deep networks, had been elusive. It's a major breakthrough. Not a "now we're safe from the Terminator" sort of breakthrough, but a breakthrough nevertheless.

It's instructive to dive a little more deeply into these already deep networks to get an appreciation for what they are doing. Which naturally brings us to the topic of BuzzFeed quizzes.

"THIS PASTA QUIZ WILL REVEAL YOUR LOVE LANGUAGE"

One activity that kept my family busy during one thousandth of 1 percent of the long summer of 2020 lockdown was a BuzzFeed quiz with the title "Say 'Yuck' or 'Yum' to These 20 Pastas and We'll Determine Your Love Language." If you haven't heard of love language, it's a pop psychology idea intended to capture the observation that different people express and expect affection differently. And if you haven't heard of a BuzzFeed quiz, you probably don't know any millennials.

BuzzFeed is a global media and technology company that got its start the same way many global media and technology companies die: by creating viral, click-baity content. BuzzFeed quizzes are one of its most popular offerings. People answer a series of questions about themselves and are rewarded with a colorful personalized assessment at the end. You can think of BuzzFeed quizzes as being like interactive horoscopes—they are *that* accurate.

These quizzes are very popular online and there are dozens of sites that host them, many of which also provide interfaces that let you build your own. And what does an online quiz have to do with machine learning and variables? I'm so glad you asked, because I was having trouble segueing to the technical stuff.

In their most common form, machine learning systems produce *classifiers*: rules that take in a description of something in terms of a set of features and then output a decision about what category the object is in. A classifier for whether a word sounds funny could use features like "Does it have a *k* in it?" and "Does it sound like a word you're not supposed to say in polite company?"

According to published research on this question—yes, the question of what makes words funny—length-adjusted unigram entropy (roughly its Scrabble score) is a highly predictive feature. That's probably why my daughter's third grade class thought the Yiddish word *"ungapatchka"* (so garish you should be ashamed at your lack of taste) was funny and adopted it as their rallying cry: "They may take our recess, but they'll never take our *ungapatchka*!"

Viewed this way, a personality quiz is exactly a classifier. The features are the answers you give to the quiz and the category is what the quiz decides you are. So, for the pasta–love language quiz, the features are things like:

- How do you feel about *cacio e pepe*? HATE it / LOVE it.
- Is this *fettuccine alfredo* making you feel things? OMG no! / OMG yes!

The five love language result categories include "Gifts," "Physical touch," and "Acts of service." Based on my answers, the quiz classified me as a "Quality time" person. Actually, that's probably about right. Damn, they're good.

It's kind of interesting that a popular internet pastime and a revolution in machine learning are related by both being classifiers. But there's even more to the connection between personality quizzes and neural networks.

When you create an online personality quiz, you have to decide how people's answers relate to the category result they will be given at the end. Typical quiz sites use a *voting* method. An answer of "HATE it" to "How do you feel about *cacio e pepe*?" could be taken as a vote *against* "gifts," a vote *for* "acts of service," and neutral on the topic of "physical touch," for example. If the quiz author associates every question with votes for and against each of the categories, then the computer can just sum up the votes at the end of the quiz and report which category got the most votes. As a quiz author, it can be hard to judge whether disliking spinach ravioli makes you more or less of a "words of affirmation" person (I'd say less), but conceptually, it's clear what you are being asked to decide.

At the end of the day, the quiz author has made a decision about the relationship between every question-answer combo and every result category. It's as if quiz authors had filled in a big table—a spreadsheet!—with one question-answer combo in each row and one category in each column. A cell contains a score for how strongly that question-answer combo in the row suggests the category result in the column, like so:

	Gifts	Physical touch	Acts of service
How do you feel about *cacio e pepe*? HATE it.	−1	+0	+1
How do you feel about *cacio e pepe*? LOVE it.	+0	+2	-½

| Is this *fettuccine alfredo* making you feel things? OMG no! | +0 | −1 | +0 |
| Is this *fettuccine alfredo* making you feel things? OMG yes! | +½ | +1 | −2 |

As a quiz author, you decide on the categories based on what the quiz is testing. Then you fill in the boxes with numbers—vote assignment numbers—which determine how the quiz will categorize people based on their answers.

When someone takes the quiz, they are selecting rows—their personal answers. The selected rows are added up columnwise, and the column with the highest score—the most votes—is the selected category.

For the table above, someone who answered "LOVE it" to *cacio e pepe* (row 2) and "OMG yes!" to *fettuccine alfredo* (row 4) would receive +0 + ½ = ½ votes for "gifts," +2 + 1 = 3 votes for "physical touch," and −½ − 2 = −2½ votes for "acts of service." Since "physical touch" has the most votes, that person would be categorized as having a physical touch love language. Don't like that choice? No problem! Just change the vote assignment numbers in the table to something that works better.

It's a little complicated, but the fact that it's part of an interface for creating personality quizzes suggests it's not rocket science. ("Rocket Scientist Quiz: Your perspective on trajectory stabilization will reveal which Disney Princess you are!")

The idea of making decisions by voting goes way back. For example, as Cleisthenes grew in power around 500 BCE, he was able to establish that the people of Athens could vote to banish budding tyrants from their midst. (Historical footnote: Shortly after, he was never heard from again. Truly.)

But a more direct predecessor of the idea we're talking about came from Frank Rosenblatt, who in the late 1950s described a scheme inspired by how brains learn to process perceptual information. He called his machine the perceptron, proving that even the earliest pioneers of machine learning knew how to pick a solid name. The essence of Rosenblatt's idea was for the machine to choose category responses by a weighted vote of its input features—precisely the same approach you find in a typical online quiz. Talk about being ahead of his time!

Rosenblatt's neurally inspired machine went further, however. His perceptron classifiers needed no human author to choose the vote assignment numbers or "weights." Instead, Rosenblatt showed that the weight the perceptron used for each feature to select each category response could be tuned automatically to match the decisions made in a collection of *examples*. That is, instead of a quiz author making up a number representing how much liking buttered noodles makes you the kind of person who craves quality time, the *machine* can choose the number grounded by the data it is given. It's a very impressive, very slick idea. Just as with the decision trees in chapter 4, instead of designing the machine by hand, Rosenblatt showed he could provide the data and the machine would design itself.

Perceptrons were the dominant model studied in machine learning, maybe even all of artificial intelligence, for about a decade. That's when Marvin Minsky, one of the founders of artificial intelligence, had had enough and decided to pour cold water on the whole overheated artificial-neurons-that-learn-from-examples-without-a-human-author craze. He wrote, with fellow MIT computer scientist and designer of the first "friendly" programming language, Seymour Papert, the book *Perceptrons*.

Perceptrons lays out a beautiful mathematical foundation for understanding the behavior of the class of computing machines Rosenblatt introduced. The authors described the choice of title as a way of honoring Rosenblatt's accomplishments, but that may have been a little disingenuous. As we'll see a bit later, they came to bury neural networks, not to praise them.

LICENSE TO NOT KILL

But first, a digression into the world of government identification cards and why computers need variables to learn about them. After it was revealed that the terrorists responsible for the attacks on September 11, 2001, had made successful use of fake ID cards, the US Congress leapt into action to close this loophole and help secure the nation. In just four short years, they created the REAL ID Act of 2005 to standardize ID cards across the country. And, only seventeen years after that, they've *almost* got it implemented.

REAL ID cards are intended to be uniform nationwide. So, naturally, the protocol for obtaining one depends completely on the state you live in. In my home state of Rhode Island, you need to provide

- one proof of identity from a list of eight acceptable documents,
- one proof of Social Security number from a list of five acceptable documents, and
- two proofs of residency from a list of nineteen acceptable documents that partially overlaps with the previous list, but even some of the overlapping documents are just a little bit different, so watch yourself.

Let's say you want to make an online quiz that tells quiz takers whether they have everything they need: "Say 'yea' or 'nay' to having these government documents and we'll reveal whether you are eligible for a REAL ID!"

Ok, definitely not the most sharable of quizzes, but it would be *kind of* useful, right? And, at the very least, it's a stand-in for other kinds of quizzes you might want to be able to write.

Here's the thing. The voting scheme we've been talking about is elegant and somewhat easy to use, but it's not *universal*. That is, there are rules, even some simple ones, that it cannot capture. And this REAL ID rule is an example. It is mathematically impossible to create a vote-based quiz that correctly determines your eligibility based on which documents you have. It can't be done. The mathematics aren't that complicated, but we don't need to go through a formal proof here. The gist is that the form of the rule creates a difficulty in using a voting scheme to tabulate the results. Think of it this way. Once you have satisfied the proof of identity, additional identity documents (no matter how many!) do not change your overall eligibility. Same for the proof of Social Security number and the proofs of residency. They are not substitutable. If you are familiar with US presidential elections, it's similar to the fact that no amount of Democratic votes in California can overturn a Republican win in Wyoming— the Electoral College is inherently a two-level system. Same thing with the REAL ID rule: it is provably and irretrievably a two-level scheme, and no single quiz can reproduce it.

The fact that there's no way to create a REAL ID quiz brings us back to perceptrons, *Perceptrons*, and variables. Minsky and Papert took three steps in their 1969 book that had a profound and negative effect on the fledgling field of neural networks. First, they referred to any classifier that uses the feature-voting scheme we've been talking about as a perceptron. Second, they described several important classes of rules—classes similar to our REAL ID example—that cannot be represented by such a perceptron. Third, they argued that, because perceptrons cannot represent some kinds of rules that any intelligent system ought to be able to represent, perceptrons are an inadequate foundation for artificial intelligence.

There are two things I want to highlight here. First of all, Minsky and Papert were correct. Simple feature-voting rules like our online quizzes are fundamentally limited. And at that time, they were the only kind of rule that could be trained by example. The book dealt a devastating blow to the field of machine learning, drastically reducing the attention—and funding—available to this line of work.

But I also want to point out that their argument, while wildly effective, was also somewhat unfair. At the time, the term perceptron was applied very broadly to a representation space of models, some of which were considerably richer than simple feature voting. It's true that researchers hadn't yet figured out an optimizer that could *train* these more complicated models, but the limitations of perceptrons weren't all that much more egregious than the known shortcomings of the programming-based approach to artificial intelligence that Minsky advocated.

One Rosenblatt supporter at the time quipped that Minsky and Papert's critique of perceptrons was "reminiscent of the mohel (person who performs ritual circumcisions) who throws the baby into the furnace, hands the father the foreskin and says, 'Here it is; but it will never amount to much.'" The piece that Minsky and Papert believed wouldn't amount to much wasn't the sum total of perceptrons.

Nonetheless, the perceptron became the circa 1999 pre-*Ironman* Robert Downey Jr. of computer science: after a string of impressive early successes, it failed to adequately deal with its personal shortcomings, and paid the price.

Perceptrons were rebranded as neural networks and had a major comeback in the mid-1980s and early 1990s. Researchers discovered that there were other ways to train neural networks from data that went beyond the perceptron training rule devised by Rosenblatt.

Let's return to the REAL ID example for a moment. There are two difficulties to overcome to create a system that can automatically learn the rule for REAL ID eligibility from examples. The first, as we discussed, is that no voting-based method can represent the rule, let alone learn it. That's a pretty big obstacle to overcome. But it's one that Rosenblatt had already worked out using variables. Here's how.

We're going to turn the REAL ID rule problem into four problems. That may not sound like progress, but the four problems are each individually much simpler.

Here's the first simpler problem. Part of the REAL ID rule is that you must provide a proof of identity by presenting any one of eight possible documents. Although there's no way to structure a quiz to express the entire REAL ID rule, what about just the "prove identity" piece? As it turns out, that's no problem.

The second part of the REAL ID rule is that you must prove your Social Security number by providing any one of five possible documents. Also a piece of cake.

The third part of the REAL ID rule is to prove your residency by providing any *two* of nineteen possible documents. That's also quite simple to do using the voting scheme. So all three subcomponents (intermediate results) can be represented as voting-scored quizzes.

Now comes the conceptual leap. Imagine a quiz where some of the answers are filled in by a *separate quiz*. Yes, we're nesting quizzes inside other quizzes. Quizception. Madness. Get on it, Buzzfeed!

By extending the game to allow nested quizzes and storing the intermediate results of each quiz in variables, we're home free; we have one quiz for proving identity, one for proving Social Security numbers, and one for proving residency. Then our fourth and final quiz takes the results of the first three quizzes and returns "eligible" for anyone who successfully proved all three of identity, Social Security number, and residency in the prior quizzes. This last quiz is also quite easy to define using a voting scheme. It just casts a one for each of the three proofs and declares "eligible" if the total is three. Cue fireworks.

By breaking down the original complicated rule into simpler rules with intermediate results stored in variables, we can use a voting scheme at each step in the calculation to build back up to the original rule. Clever. And again, Rosenblatt knew all this.

Unfortunately, Rosenblatt's learning rule for perceptrons doesn't work for these multilayered perceptrons that correspond to our nested quiz idea. You can get it to work, but you need all the intermediate results ("identity proven," "residency not proven," etc.) to be available, along with the examples being used for training. That is, for the machine to work out a complex rule, someone must have already decomposed it into simpler rules. That's cheating.

So the solution to problem 1—nest quizzes inside other quizzes and store their results in variables to represent complex rules—has left us face-to-face with problem 2: How do you train such a scheme from examples?

Here's the trick. You can define a loss function as discussed in chapter 3 and chapter 4 that scores the vote assignment numbers for a quiz—really, any neural network—based on how well it classifies a set of training examples.

Recall that the machine learning recipe requires three ingredients: a representational space of programs to consider, a loss function for scoring the potential programs, and an optimizer for searching for a program with a good score. The representational space we're talking about now is the vote assignment numbers that define how the quiz will work.

To compute the loss for a given set of vote assignment numbers, run the training examples through the quiz. Examples for which the vote assignment numbers produce desired answers incur no penalty, but examples that produce incorrect answers are penalized in proportion to how badly the correct answers were outvoted. You can forgive the country for choosing a bad president by a small margin, but electing a terrible president in a landslide, that stings. Ouch.

The reason this kind of loss function is so useful is that it makes it possible for the optimizer to play a digital version of the "hot and cold" game. The computer can tell whether small changes to the vote assignment numbers bring it closer to the target rule or move it further away. The optimizer works out, for each possible small change to a vote assignment number, whether the overall set of vote assignment numbers produces decisions more like the desired rule ("You're getting warmer!"). Behind the scenes, making it possible to connect the vote assignment number changes to the loss function changes—thereby providing hints for how to improve the vote assignment numbers—is a fancy bit of mathematics called calculus. Yeah, that calculus. The one many of us learned some-what and couldn't possibly imagine ever caring about. It might not have helped you much in your career, but calculus sure is making computers a lot smarter. That's because calculus calculates slopes of functions, which is precisely the question of how a small change in the x-axis input (a vote assignment number) impacts the y-axis output (the accuracy of the rule).

When researchers in the 1980s hit on this idea of training multilayer perceptrons using calculus, it sparked a revolution. Researchers through-out engineering and the sciences jumped on the bandwagon to find ways of applying these neural networks in their work.

But, like Robert Downey Jr.'s successful comeback at the turn of mil-lennium, the happy times didn't last. By the mid-1990s it was becom-ing clear that neural networks required so much data and computer power that there weren't many problems that were well suited to their strengths. The machine learning community's attention drifted to other approaches, including decision trees and a souped-up version of the old voting approach called support vector machines. (Score another nam-ing win for machine learning researchers.) Neural networks went deep underground.

The Google-computers-watch-cat-videos moment brought them back into the light. With phenomenal increases in computer speed, the avail-ability of mind-warping amounts of data, and some new algorithmic innovations, deep networks started racking up impressive accomplish-ments. Problems that computer scientists had been chipping away at for decades succumbed to the deep learning tsunami. Want a program that can tell the difference between Bedlington, Border, and Boston Terriers? Check. Want a program that can transcribe what you are saying? You got it. Want a program that can outflank the best human beings in the ancient strategy game of go? Child's play.

So neural networks are back yet again. They are a huge commercial success and have become synonymous with futuristic, cutting-edge tech-nology that is having an impact on our world. This time around, it feels as though they are back to stay. Kind of like Tony Stark himself, Robert Downey Jr.

6

CONSOLIDATING INTO LOOPS
YOU DON'T HAVE TO TELL ME TWICE

Whether it's in your hairstyle, your model airplane flying, or your cereal bowl, loops make an exciting addition. Your vocabulary for conveying tasks to a computer is no different.

A loop is a set of instructions or behaviors that is repeated multiple times. The most obvious advantage of communicating tasks using loops is that you can say a lot with very few words. Imagine you are in gym class and the instructor says you need to do a pushup, a pushup, a pushup, a pushup, a pushup, rest for a moment, then do a pushup, a pushup, a pushup, a pushup, a pushup, rest for a moment, then do a pushup, a pushup, a pushup, a pushup, a pushup, and stop. Phew! The effort it took to listen to the instructions left you too tired to actually do the exercise. A more parsimonious way to express this sequence would be for the trainer to say "Do three sets of five pushups." As the number of sets and reps (repetitions without a break) grows, the number of saved words grows *exponentially*.

Being able to convey a long activity in very few words is nice in and of itself, but there's a bunch of other advantages that come along with it. The shorter sequence is easier to remember, easier to follow, and easier to modify. It calls out the common structure in the sequence, which makes it easier to carry out the activity with less attention to the details.

Ikea, the Swedish furniture company famous for do-it-yourself assembly and meatballs, puts a lot of effort into flagging repetitions in its instructions. In the manual for its Hemnes eight-drawer dresser, it show a picture of the reverse of the piece of furniture with a whole bunch of nails standing at the ready to hold the backing board in place. But how do we secure these nails? Ikea provides a little diagram of a hammer hitting a single one of the nails into place. Should we just hammer in that one nail? No, the hammer is shown in a little speech bubble, which carries the notation "28x." It's telling us that we should repeat this hammering action for all twenty-eight of the nails shown. If Ikea hadn't provided the "28x," it might not be clear that this action should be repeated. But if it had drawn twenty-eight hammers on the diagram, it would have been cluttered and would oblige the customer to look at each of the twenty-eight pictures to see if there was some important difference. Not expressing the looping behavior explicitly makes a complicated process much, much worse.

As with the sets-reps example above, Ikea sometimes makes use of *nesting*, loops within loops. To walk us through attaching the drawer pulls on the Hemnes, Ikea shows a picture of a screwdriver twisting a screw into the knob to link them together. As in the nail example, the screwdriver-knob-screw group has a bubble around it with a "2x" label, indicating we need to repeat that action for the two knobs on the drawer. Above this picture is a diagram of a deep drawer with a "2x" on it and a diagram of a shallow drawer with a "2x" on it. They are in a box with an arrow pointing to the knob-assembly step, suggesting that we will repeat that step for each of the drawers in the box. That means we'll repeat the two-knob assembly for each of two deep drawers and two shallow drawers. How many knobs are we assembling? 2x (2x + 2x) = 8x! One screw picture is ultimately applied to eight different parts of the dresser. All told, the rather harmless Hemnes chest of drawers requires forty-eight—FORTY EIGHT!—pages of instructions. I estimate that, without the repetition notation, the instructions would be . . . well, a lot longer and much more confusing.

Sets-reps in exercise plans and call-out bubbles in Ikea instructions represent one of the simplest kinds of loops we'll look at in this chapter, a *counting* loop. Counting loops are a special case of a more general kind of loop applied to all the elements of a set. In counting loops, the elements of the set are specific cardinal numbers: Often we count silently as we do the activity, associating a number with each repetition. But sometimes the set consists of a collection of specific objects. A helpful website gives these instructions for shucking corn:

1. Peel off the outer leaves.
2. Expose the tip of the ear of corn.

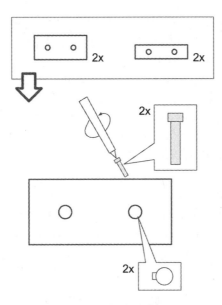

Figure 6.1 Ikea uses nested loop diagrams that allow one picture of a screwdriver to represent fastening eight knobs.

3. Grasp the tops of the leaves and the tassel.
4. Pull down in one firm tug.
5. Break off the leaves and the silks.
6. Tidy up the cob.
7. Repeat with the remaining ears of corn.

That last step is the interesting one right now. We are to repeat this sequence of steps *for each* ear of corn. How many ears of corn? Well, that doesn't really matter—we can use these instructions regardless of the size of that set because they are just telling us that each ear in our basket needs to be processed.

Programmers sometimes call this construction a *for each* loop, as in "for each ear of corn . . . ," since we're acting on each of the items in the set. These *for each* loops, in turn, can be seen as a special case of an even *more* general looping structure.

Let's look a little more closely at step 2 above. How do we expose the top of the ear of corn? The online instructions read "Peel back the leaves at the tip of the cob just until you can see the top few rows of kernels."

That is, we start peeling and then keep on going, stopping only once we can see some of the corn kernels. Pull a little, look for the kernels. Pull a little more, look for the kernels again. We don't know how many times we're going to need to pull. It's not by count or by inches or by every member in a collection. It's something we need to repeatedly check. It's a *condition*, just like what we talked about in chapter 4, but one we check repeatedly. While the condition remains true, keep on shucking.

For this reason, this kind of loop is often called a *while* loop. We can formulate a counting loop as a *while* loop (while we haven't reached the target count, keep lifting those kettlebells). Same thing for *for each* (while there are still ears of corn we haven't peeled, keep peeling). In the next chapter we'll see an even more general way to construct loops. But I think these three will keep us busy, at least for a while.

IT'S GROUNDHOG DAY ALL OVER AGAIN

If you ever doubt the centrality of loops in modern life, flip through a calendar. A typical appointment book has four fifteen-minute slots each hour, eight hours each day, five workdays each week, around four or five weeks each month, and twelve months each year, year after year after year. We break up time into loops upon loops.

Back when we used to keep our appointments on paper, we'd have to copy repeating events each week (dance lessons), each month (rent due), and each year (Cousin Judy's birthday). Now calendar apps make it (relatively) easy to keep our appointments organized and we can create rules—little programs, really—that keep our recurring events straight.

As a concrete example, let's make a calendar event for Juneteenth, which is observed in the US every year on June 19 to commemorate the emancipation of enslaved African Americans. (I used Google Calendar for this example, but I've also tried it in Apple's Calendar application and Outlook's calendar, both of which are similar.) Creating the event on June 19, 2022, provided me with a set of options regarding how the event would repeat:

- Does not repeat: It's just a one-time thing.
- Daily: It happens every day starting June 19, 2022.

- Weekly on Sunday: Google Calendar notices that June 19 falls on a Sunday in 2022 and provides the option of repeating the event that day every week.
- Monthly on the third Sunday: Google Calendar notices that the Sunday it falls on is the third one in the month and suggests repeating this pattern.
- Annually on June 19: Right; that's what I was going for.
- Every weekday (Monday through Friday): Considering that June 19, 2022, is a Sunday, that's a pretty weak guess, Google. I expected better.

That's an impressive set of choices, but there are other ways the event might repeat, of course. It might be every third day starting June 19, 2022. It could be each month on the nineteenth day of the month. It could be the second-to-last Sunday of the month. All of these patterns are consistent with the selected date. Understanding and expressing *how* a pattern repeats is one of the things that makes loops so powerful and also so challenging. After all, if you get the pattern right, the computer can mechanically fill in the event *everywhere* it applies; you never have to think about it again. But the calendar program can't properly generalize from a single example—not with any confidence, anyway. Philosophers call that the problem of induction. At least I think that's what they call it. I heard them talking about it once and I just assumed that was the general principle they had in mind.

Google Calendar lets you pick from this small set of patterns because the programmers assumed this set contains the most common ones. The longer the list, the more confusing the interface. The shorter the list, the more likely it is that the pattern you need won't be on it. It's a balancing act.

Anyhow, the interface hedges against the possibility of missing your particular pattern by also providing one last choice, Custom. In Custom mode, you can choose the frequency (daily, weekly, monthly, yearly), how many of these units to skip (never skip, skip one, skip two, etc.), and when the pattern ends (on a specific date or after a specific number of repetitions). In addition, if you select "weekly," you can pick which days of the week to use (only Mondays, both Fridays and Saturdays, etc.). If you pick monthly, you can choose the specific day of the month (the nineteenth) or the selected weekday (such as the third Sunday). That's a lot of configurability! It handles two out of the three examples above of patterns not covered by the original menu of choices. But it doesn't let you pick "second-to-last Sunday of the month." Maybe that's okay. After all, short of allowing specifications of arbitrary formulas, there always will be some patterns that are left out.

After a little poking around, I discovered these calendars do not support arbitrary formulas. But they do support a richer set of choices than what you can find in the interface. Back in 1998, the big tech companies of the day (remember Lotus?) were starting to notice that: (1) people were finding it useful to keep their calendars organized digitally and (2) people who created their calendars in some other company's software would have to toss it and start from scratch if they wanted to switch to your software—something people tended not to want to do. The companies concluded that a standardized format for calendar entries—including repeating events and a bunch of other bells and whistles, such as whether you wanted bells or whistles to alert you before the event—would be very nice. They wrote up a standard with the sexy official name RFC 2445, which, these days, also works as a celebrity baby name. (RFC stands for "request for comment." It's how all internet standards are made public. I'm not sure the authors *really* want comments, but that's what the documents are called. Regardless, the world is a better place because of the people creating these standards. We salute you.) The official name of the standard is "iCalendar," for "internet calendar."

Most calendar programs today read and write files in iCalendar format. The way they represent repeating events internally is a little arcane but fascinating. As I said, internet calendars can do more than most calendar programs reveal through their interfaces. Let's take a look at the piece of the iCalendar format that deals with repeating events—loops.

Here's how to express Juneteenth:

```
RRULE:FREQ=YEARLY;BYMONTH=6;BYMONTHDAY=19
```

It's not super user-friendly, but we can take it apart piece by piece. RRULE means "recurrence rule." That is, it's announcing "here's the part of the calendar entry that covers how the event repeats." FREQ=YEARLY says we're talking about something that happens with a frequency of once a year. BYMONTH=6 says that it happens in the sixth month, June. BYMONTHDAY=19 conveys which day of the month: the nineteenth. Not too terrible, right? It's saying that the event "repeats every year, in June, on the nineteenth." Other events that happen on a specific date each year are expressed similarly. Halloween is:

RRULE:FREQ=YEARLY;BYMONTH=10;BYMONTHDAY=31

and my birthday is

RRULE:FREQ=YEARLY;BYMONTH=8;BYMONTHDAY=30

in case you want to send me a card.

If we get a paycheck on the fifteenth of each month and want to mark our calendar accordingly, we can modify the frequency and schedule it as:

RRULE:FREQ=MONTHLY;BYMONTHDAY=15

It says that the frequency is monthly and the "monthday" it repeats on is the fifteenth.

These examples are easy to express in Google Calendar. But here's a cute trick that Google Calendar doesn't expose in its interface. Even though we are saying the event is monthly, we can actually specify a set of days instead of just a single day. If we get paid on the fifteenth and the thirtieth, we could include a list and write:

RRULE:FREQ=MONTHLY;BYMONTHDAY=15,30

That rule creates two events each month. But it's not quite right, is it? Some months don't even have a thirtieth day. We actually get paid on the fifteenth day and the *last* day of the month. How can we say that? iCalendar's got you covered:

RRULE:FREQ=MONTHLY;BYMONTHDAY=15,-1

The −1 here means "last." You can write any negative number and it will count backward from the end.

What if you get paid on the first *workday* of the month? It's not always the first and it's not always a Monday (although Monday is three times more likely than any other workday to be the first workday of the month). iCalendar has a slick feature that makes it possible to express this idea. As we saw in the payday example, we can use lists to specify a *set* of events to repeat. Well, once you've created that set, you can tell iCalendar to restrict the events to a particular position in this set. The command

```
RRULE:FREQ=MONTHLY;BYDAY=MO,TU,WE,TH,FR;BYSETPOS=1
```

first creates an event on all the workdays of the month. The BYDAY part of the command lists the days of interest—workdays only. The BYSETPOS=1 part of the command says "of all the elements in this set, just use the first one." That is, of all the workdays in the month, schedule this event on the first such day. That's a neat concept, and I've never seen a calendar program that lets you make use of it, even though they all understand it.

Thanksgiving in Canada takes place every year on the second Monday in October. We could express this pattern using BYSETPOS by activating all the Mondays and selecting the second one. But since this construct is really common in calendars, iCalendar allows for a shorthand—a number before the day of the week, as in:

```
RRULE:FREQ=YEARLY;BYMONTH=10;BYDAY=2MO
```

So the day is specifically 2MO, the second Monday. Coincidentally, we've got only 2MO examples left.

Combining a few of these ideas, we can express Father's Day in the Dominican Republic, which is celebrated on the last Sunday in July:

```
RRULE:FREQ=YEARLY;BYMONTH=7;BYDAY=-1SU
```

I could go on all day (every third Thursday of a month that begins with A), but my point isn't to provide an exhaustive introduction to iCalendar but to get across three important ideas. First, the concept of loops is pretty useful—literally something you can use every day. Second, iCalendar provides a well-thought-out specialized language for expressing the kinds of loops commonly used in calendars. Finally, much of its power is hidden from us by the interfaces we use to interact with them. That last point frustrates me. I'm all in favor of making the easy things easy, but can't we make the complex things at least possible? If we hide from this kind of complexity, system designers will inevitably hide its power from us.

I can't resist presenting one more example. I was wondering whether iCalendar could help people born on February 29 mark their birthdays.

After all, if you make a yearly event on February 29, it will only show up on the calendar on leap years. Anyone who has ever seen *The Pirates of Penzance* knows that observing one's birthday only in leap years can lead to heartbreak and musical mayhem. (Sorry, Frederick, you'll need to wait sixty-seven more years until your twenty-first birthday.) On the other hand, celebrating each year on February 28 seems defeatist. Can we ask iCalendar to put the event on February 29 if there is one and February 28 when there is no other option? Here's what I came up with. My solution combines a bunch of ideas we've talked about all in one:

```
RRULE:FREQ=YEARLY;BYMONTH=2;BYMONTHDAY=28,29;BYSETPOS=-1
```

That is, every February, we make a set of dates consisting of the twenty-eighth and the twenty-ninth. In leap years, we get both of these dates, and in other years we get only the twenty-eighth (because the twenty-ninth doesn't exist). Then we use BYSETPOS to select one of these days, specifically the *last* one (−1). In leap years, that's the twenty-ninth. In years with no twenty-ninth day of February, that ends up being the twenty-eighth. Boom! Frederick gets his yearly birthday after all. I'm just glad Gilbert and Sullivan finished their operetta before RFC 2445 came out, or Act 2 of the show would have been a whole lot less entertaining. (When my daughter/humor-writing consultant Molly read this chapter, she pointed out that the "last day of the month" rule would have been a simpler solution to this problem. As I said, she's clever!)

IT'S ALL FUN AND GAMES, FOREVER

Coming back to the different types of loops, the *while* loop checks its condition each time around the loop. If the condition is true, the loop continues. Let's think of what would happen if the condition was the logical expression "true." Or, if that seems too weird, a truism, such as "1 + 1 = 2." Such a loop would never complete because the condition would never stop being true.

Loops that never complete are a bogeyman in computing. They are often referred to as "infinite loops." They are like ominous black holes from which no program will ever escape. The question of whether a given program will be captured in an infinite loop is of both practical and theoretical importance. On the practical side, we ask ourselves this question regularly. My computer has become nonresponsive. Is it just a little hiccup, or is it permanently stuck? If it's the former, I want to wait a little longer to let it come back. If it's the latter, I might have to try the old turn-it-off-and-turn-it-back-on-again solution. Being able to accurately tell whether a program has fallen into an infinite loop would be useful for developers looking for bugs or for the rest of us trying to make sure that valuable work isn't lost by restarting a computer unnecessarily.

On the theoretical side, the question of whether a program is doomed to be stuck forever or, alternatively, that it will complete its work and halt is referred to as the "halting problem." Computer science pioneer Alan Turing in the 1930s showed that solving the halting problem in complete generality—correctly recognizing all possible infinite loops—is impossible; no computer program can do it. These days, computers are so fast and so powerful that it may not be so obvious to people that there are well-defined mathematical problems no computer can solve no matter how much time it is given. It's kind of humbling. Could a computer scientist devise a problem so complex that no computer scientist could solve it? As it turns out, yes. Turing did it.

Yet in all the worrying about infinite loops and halting problems, sometimes people forget that some of the most important programs we run are based on infinite loops. A computer operating system is itself a program that is ever vigilant, monitoring for clicks and disk access and managing other running programs. We want the operating system to keep running indefinitely and never halt—after all, halting in this case would be crashing. More generally, any computational device, be it a cell phone, a video arcade machine, a climate control system, a network router, or a car, is tasked with "keep doing what you are supposed to be doing as long as you possibly can."

Even systems that don't seem to have loops in them often have an implicit *do forever* wrapped around them. In chapter 4, we looked at trigger-action programs. A trigger-action program is just a collection of rules like the following:

- **if** the sun goes down: <u>turn on</u> the porch lights
- **if** the clock strikes midnight: <u>turn off</u> the porch lights

Here, "sun goes down" and "clock strikes midnight" are triggers and "turn off the porch lights" and "turn on the porch lights" are actions. Although there is no explicit looping noted in the program, some kind of *do forever* loop has to be used here. After all, if we execute each of these rules just once, sometime around 2 p.m., say, nothing will happen. The way we should think of a trigger-action program consisting of a set of rules is more like this nested loop:

```
do forever:
    for each rule r in our trigger-action program:
        if r's trigger applies: execute r's action
```

Sneaky, huh? We've kind of been talking about loops all along. So let's double down. We can create behaviors that are more explicitly "loopy" with an extension of the trigger-action paradigm. This extension is commonly used in the building video games.

Video-game "engines" provide a useful programming environment for creating games. There are dozens of them out there these days, with some focused on helping people build commercial-quality independent games and others designed with learning and ease of use in mind. Video-game companies have their own proprietary engines, which are seen as their competitive edge in creating experiences no one else can produce. Despite the diversity, video-game engines generally support a kind of programming that extends trigger-action programming.

Let's talk through, at an intermediate level of detail, how we could build a specific game. Oh, I have an idea for a game! It's inspired by an experience I had walking the family dog, Sadie, after a snowstorm. Because the roads had been plowed, they were walkable, but lined by tall snowbanks that prevented the dog from leaving the road. It was trash day, so there were lots of trash cans along the road that Sadie wanted to sniff. But it also meant there was a trash truck trundling through the streets, which she finds very, very frightening. Usually, Sadie runs onto the sidewalk to get away if a truck gets close, but the snowbanks blocked her that day. So the stakes were high. She wanted to visit all the trash cans but never get too close to the trash truck. It felt like a real-life version of a famous arcade game from the 1980s. We'll call my version of the game Pac Dog to avoid any possible intellectual property concerns. The dog moves around in a simple neighborhood street grid. If she sniffs each trash can once, she wins the game. If she gets too close to the trash truck driving around the outskirts of the neighborhood, she loses the game. (No dogs were harmed while designing this game. But she did get quite scared at one point.) We'll specify the game as a set of trigger-action rules.

First, here's how game control works:

- **if** left button pressed: <u>move dog</u> left <u>one step</u>
- **if** right button pressed: <u>move dog</u> right <u>one step</u>
- **if** up button pressed: <u>move dog</u> up <u>one step</u>
- **if** down button pressed: <u>move dog</u> down <u>one step</u>

These four rules are all we need to get the dog moving around the streets. But we need to add a rule to keep the dog from going through snowbanks:

- **if** *dog* collides with *snowbank*: <u>bounce dog</u> back <u>one step</u>

Something special happens if the dog runs into (sniffs) a trash can. That trash can is removed from the game, and the dog's score goes up. If all the trash cans are gone, the dog wins the game.

- **if** *dog* collides with *trash can*:
  <u>change score by adding</u> 10 points
  <u>destroy</u> *trash can*
- **if** the number of *trash cans* equals zero:
  <u>game won</u>

Let's get the trash truck into the action. If the dog collides with the trash truck (or vice versa), the dog loses:

- **if** *dog* collides with *trash truck*:
  <u>destroy</u> *dog*
- **if** *trash truck* collides with *dog*:
  <u>destroy</u> *dog*
- **if** the number of *dogs* equals zero:
  <u>game lost</u>

That's all we need to do to specify the core dynamics of the player. In many ways it's not that different from the home automation examples we went through in chapter 4.

The last piece I want to talk about is handling how the trash truck moves. To keep things simple, let's just have the trash truck move in a simple box pattern. It moves to the right until it reaches the end of the road. Then it moves down until it reaches the end of the road. Then it moves left until it reaches the end of the road. Then it moves up until it reaches the end of the road. In the map I'm imagining, that sequence brings the trash truck back to where it began, and we can just have it repeat this pattern indefinitely.

Using the *do forever* and *while* looping structures we've talked about, we can specify this behavior pretty easily.

```
do forever:
    while trash truck is not colliding with snowbank:
        move trash truck right one step
    bounce trash truck back one step
    while trash truck is not colliding with snowbank:
        move trash truck down one step
    bounce trash truck back one step
    while trash truck is not colliding with snowbank:
        move trash truck left one step
    bounce trash truck back one step
    while trash truck is not colliding with snowbank:
        move trash truck up one step
    bounce trash truck back one step
```

That seems pretty reasonable, and it's a nice demonstration of two of the looping structures we covered earlier. But unfortunately, there's a problem. In the trigger-action programming style, we can't have loops within loops. There is just one loop—the implicit outermost *do forever* loop. One loop to rule them all, as they say. We don't want other loops to be nested inside this loop. After all, that would result in everything else in the game stopping while the trash truck just went around and around and around forever. Instead, we want the trash truck's behavior loop to take place simultaneously with the movement of the dog and other aspects of the game.

Turning the doubly nested looping structure defining the trash truck's behavior into a set of trigger-action rules requires a new way to use a concept we talked about in chapter 4: *state*. State is the context of an ongoing process that's necessary to know what will happen next. You can think of state as what a robot would need to remember so that, if it were to be interrupted in the middle of carrying out its task, it would be able to pick up later from where it left off. The Polaroids and tattoos used by the protagonist in the movie *Memento* are his state so that he can continue to carry out an investigation despite not being able to form new memories. I use my calendar appointment list and to-do list this way to make it through the day despite too often embodying the stereotype of the absent-minded professor. I live in fear that my students will discover the lack of security on these important lists and will realize they can send me to Poughkeepsie to yodel at midnight by just adding the right entries to the right lists.

Where were we? Oh, yeah, state. We can decompose the behavior of the truck into four states: going right, going down, going left, and going up. They correspond to which of the four directions the truck could be going. We'll use a variable, as in chapter 5, to record the trash truck's current state, and we'll use its four values—rightgoing, leftgoing, upgoing, and downgoing—in our trigger-action rules for directing its behavior. First we translate the state into movement for the trash truck, much as the button presses were translated into movement for the dog:

- **if** *trash truck state* is rightgoing: <u>move</u> *trash truck* right <u>one step</u>
- **if** *trash truck state* is leftgoing: <u>move</u> *trash truck* left <u>one step</u>
- **if** *trash truck state* is upgoing: <u>move</u> *trash truck* up <u>one step</u>
- **if** *trash truck state* is downgoing: <u>move</u> *trash truck* down <u>one step</u>

Next we need to update the state, depending on what happens in the world. For the trash truck behavior, the state changes when the trash truck encounters a snowbank. Specifically,

- **if** *trash truck* collides with *snowbank*
    AND *trash truck state* is downgoing:
    <u>bounce</u> *trash truck* back <u>one step</u>
    <u>change</u> *trash truck state* to leftgoing
- **if** *trash truck* collides with snowbank
    AND *trash truck state* is upgoing:
    <u>bounce</u> *trash truck* back <u>one step</u>
    <u>change</u> *trash truck state* to rightgoing
- **if** *trash truck* collides with *snowbank*
    AND *trash truck state* is leftgoing:
    <u>bounce</u> *trash truck* back <u>one step</u>
    <u>change</u> *trash truck state* to upgoing

```
· if trash truck collides with snowbank
     AND trash truck state is rightgoing:
     bounce trash truck back one step
     change trash truck state to downgoing
```

To make sense of this program, let's compare its structure to the original specification of the trash truck's behavior. Before, the movement of the trash truck was written with a collection of four *while* loops. Each one moved the truck in a direction until it hit a snowbank, then progressed to the next *while* loop. This new structure captures the same idea by using four states. Each state corresponds to one of the original *while* loops. When a given state is active, the trash truck moves in the direction that the corresponding *while* loop used. When the condition in a *while* loop is no longer satisfied, the new structure changes the state to correspond to the next *while* loop in the sequence. In this way the original *while* loops are folded into the implicit *do forever* loop of the video-game engine.

By breaking up the trash truck's loops in this way, its behavior is now effectively simultaneous with the dog's, and the two occupy the same reactive environment. Wanna try to beat my high score?

Infinite loops can definitely cause trouble. Blindly following the shampooing instructions "lather, rinse, repeat" is one of the leading causes of drowning among beginning programmers. But *do forever* loops are both necessary and sufficient for computation. They are sufficient because we can translate other kinds of loops into this form using the concept of state. And they are necessary because any computer system should be trying to keep itself healthy and working for as long as it can.

LOOPS OF LEARNING

As we saw in the previous chapters, the power of machine learning is that it can help express tasks that are too difficult to author explicitly. The same is true in the context of loops. A supermarket shopping robot might be asked to buy a lemon from the produce section. As a loop, the chore looks like:

```
for each area in the produce section:
    look at the fruit or vegetable in the area
```

```
if the fruit or vegetable is a lemon:
    pick up the fruit or vegetable
    put the fruit or vegetable in the shopping cart
    exit the loop
```

There are a few places here where machine learning could be a big help. If all areas have labels written in English (or, even better, a QR code or a UPC code), the robot can tell whether it has found the lemons. If not, being able to recognize a lemon visually is a perfect fit for a machine learning–created deep network image classifier. (It's good to know a lemon when you see one.) Accurately picking up one of the lemons without squishing it or knocking it to the ground is also challenging, and the best programs for choosing and executing a grip using a robotic hand are also machine learning based. In both cases, we train the decisions by giving lots of labeled examples—lemon/not lemon, good grip/ bad grip. The combination of a "classic" loop with a machine learning recognizer inside it is a great way of getting the task across to the robot shopper. Hmmm, wait a sec. I need to put "write robot shopper tv pilot" on my to-do list.

In this example, we know we want a lemon and not a box of cereal. What if we wanted the robot to handle a higher-level task? Let's say we want it to do the week's shopping. We could train it by doing the shopping ourselves once and then having the robot copy us. The problem is that what we want copied isn't that the robot should repeat the same purchases we selected in our demonstration. After all, that toothpaste we got was necessary last week, but we don't need a new tube every week. To correctly identify products to purchase, the robot needs more context than just "is that a lemon?" It needs to remember that we're running low on toilet paper or bagels or whatever.

A straightforward way of applying the machine learning approach we've discussed to this situation is to build a data set of context plus product, paired with whether that product should be purchased. Here, context would refer to all the various things the robot saw and heard during the week, like Alexa in always-on mode. Given enough examples, a machine learning algorithm should be able to learn to tell whether a product is needed based on the available context. But it might take a lot of examples. Is the fact that your mother is visiting relevant? (Maybe, if you want to have her favorite brand of beverage available. Maybe not if she always brings her own because she doesn't trust you to get it and still treats you like a little kid and simply won't admit that you are a grown man now and you can take care of other people, including the two kids you raised, and why can't she ever let me live down that time I forgot to pick her up from the train station twenty years ago?) Oops, sorry.

Anyhow, if the learning algorithm considers too little context, it won't have enough information to do the right thing. If it considers too much context, it will be overwhelmed by details, and learning will become too hard. One solution to this dilemma is to provide more than just examples. We can tell the robot, "I like to eat berries as a snack at night. You should make sure I have them, but don't be wasteful." Instead of just an example, you provide a justification as well. That additional guidance makes the learning problem much more feasible.

The US Supreme Court does a similar thing. It's too much to ask everyone to figure out what the court thinks is constitutional or not based on examples. Sometimes it seems like the court is all over the place. So, when the justices issue a ruling, they provide a justification. For example, the court ruled in 2018 that the government violated the Fourth Amendment to the US Constitution (no "unreasonable searches and seizures") by accessing location information from cell phone records without a search warrant. That outcome could be seen as surprising because previously, the court had ruled that when people sign up for a service (such as cell phone service or a bank account or a gym membership), they are volunteering their information and police can get such information without a search warrant. Maybe the court no longer believes that's true? From the decision alone, there's no way to know. But, along with the decision, they spelled out what caused them to rule this way. Voluntarily shared information is still fair game. But location information on cell phones is a special case because it's just too good. In their published opinion on the case, the justices wrote that "when the Government tracks the location of a cell phone it achieves near perfect surveillance, as if it had attached an ankle monitor to the phone's user." And that's a step too far to allow warrantless access.

The justification fills in a lot of the gaps that are left by looking just at the examples themselves. The same idea is beneficial in machine learning. Instead of just giving a learning algorithm an example of the kind of behavior that should be executed in a loop, we can provide a description of what the loop is supposed to *accomplish*, giving the learner a way to assess its performance autonomously and learn far more from far fewer examples. In terms of chapter 1's 2 × 2 grid, we're talking about explaining, aka explicit incentives, instead of demonstrating, aka example instructions.

Returning to the Pac Dog example from earlier, let's imagine that we want to tell the dog what to do so it can win the game. We worked through how to lay out the game itself, and provided behavior for the trash trucks. But selecting and specifying a strategy for the dog is considerably harder. A subarea of machine learning called *reinforcement learning* is about using experience to fill in the decisions inside a loop like this.

We provide three things to a reinforcement learner so it can do its job: (1) a set of commands or actions the learner can take in the loop. They are akin to the actions in trigger-action programming; (2) a set of features of the context that the learner can use to decide what action to choose at any given moment. These features are akin to the states in trigger-action programming and are usually referred to as *state features*; and (3) a description of the objective so that the learner can evaluate its own performance and try to find ways to behave better. There's no direct analogy to trigger-action programming for the third part. Having an explicit objective is the main distinguishing feature of reinforcement learning.

For Pac Dog, the actions are the four movement commands the dog can take at each step. The state features are things like where the dog is, where the trash truck is, and which trash cans are still left to sniff. The objective is to sniff all the trash cans without contacting the trash truck. The beauty of the reinforcement learning approach is that we're kind of done at this point. Once we pick the actions, state features, and objective, it's up to the learner to figure out how to behave to achieve the objective. I say "kind of done" because it's pretty easy to choose an objective that is too hard for a learner to achieve or one that, even if achieved, doesn't actually do what you want. So I guess by "kind of done," I really meant "just getting started." So let's look a little more closely at objectives.

HOW DO YOU MEASURE A YEAR?

At the heart of the idea of reinforcement learning is evaluation. The learner is not told what to do. Instead, the learner is given a way to score itself and must figure out how to change its behavior to increase its score. It's a lot like the optimizer minimizing loss in the machine learning recipe of chapter 3 except that it may need to act in the real world to find out how it's doing. Reinforcement learning is a lovely idea because it encourages good solutions without having to spell out in advance how to make them happen.

Of course, you don't get something for nothing. There's still a matter of choosing the details of how the evaluation will be applied. Any solution to a problem involves trade-offs—costs and benefits. For reinforcement learning to work, these costs and benefits must be expressed in terms of a common currency so that the overall solution can be assigned a score. One of the most natural forms of currency is, well, currency: money. A reinforcement learning–based stock trading system, for example, has costs in terms of the price of the stocks it buys and benefits in terms of the price of the stocks it sells. In both cases the outcome of the decisions can be expressed in dollars. That's great because it provides an objective way for the system to tell how well it's doing. Money is a good choice because other quantities can also be expressed in terms of dollars. For example, time is money.

Although, for that matter, time is time. Time is another unit of exchange for decision-making. Should a navigation system send you on the highway or local roads? Well, which will get you there faster?

My PhD student Dave Abel put together a team that experimented with another interesting evaluation measure. They built a physical solar panel that could angle itself in different directions. Moving the panel to point at the sun could bring in more energy, but it also had a cost in terms of the energy needed to drive the motors. So, in that project, the common currency was electrical current (see?).

Evaluating and rewarding performance is central to sports competitions. Swimmers and runners are measured in time (faster is better). Jumpers and throwers are measured in distance (farther is better). Team sports typically have a scheme for scoring points, with the team amassing the most points at the end of the bout being declared the winner.

The common currency issue is of central concern in how a sport is designed. Consider the rules in basketball. The winner is based on who has the most points, which seems pretty straightforward. But there are subtleties. For example, there are a few ways to score points, and a decision needs to be made about points' relative worth. A successful free throw is worth one point. Scoring a basket in regular play (a field goal) is worth twice as much. And, if the field goal is taken from far enough away, it's worth three times a free throw. The choice of assignment from outcome to points has an impact on the game strategy. If a field goal were worth ten times as much or one-hundredth as much as a free throw, the incentives for how to play would be different.

In addition to how valuable it is to score in different ways, it's also important to quantify how bad it is to violate various prohibitions. Basketball players are not supposed to do certain things, such as elbowing an opposing player in the nose or running with the ball without dribbling. To discourage these behaviors, there has to be a consequence, and that consequence is meaningful only if it has an impact on the team's chance of winning the game. A rule infraction in basketball can result in a range of penalties. Some give the other team a chance to take a free throw. Or two. Or two, but only if the first attempt is successful. Or even three in some cases. Sometimes they even throw in a pizza party, but that's not an official part of the rules. Some penalties cause control of the ball to move to the other team or the person who committed the foul to be removed from the game. The existence of these various penalties do not stop teams from committing fouls. But they do determine how costly fouls are in terms of their impact on winning or losing. A coach will sometimes instruct a player to commit a foul intentionally if the benefit of doing so is judged to exceed the cost.

The common currency issue is particularly acute in the heptathlon. There, athletes compete in a diverse set of seven individual track and field events, such as the 200-meter dash, the shot put, and the high jump. The Olympic motto translates to faster, higher, stronger. If one athlete is able to outperform all others in all seven events by running faster, jumping higher, and throwing stronger, it's easy to pick a winner. But what if one competitor jumps an inch higher, but runs five seconds slower? The answer has changed over the years, but the current system is an interesting case study in designing explicit incentives. It goes like this:

Each event has a benchmark value that the athlete must achieve to earn points in that event. For example, the high jump benchmark value is 75 cm (about 2.5 feet) and the 200-meter dash benchmark value is 42.5 seconds. The more that the athlete improves over this benchmark value (by jumping higher or finishing faster), the better. To reward improvements at the extremes, this difference is rescaled by raising the difference to an event-specific power greater than one. For the high jump, that power is 1.348. Then this total is scaled into a reasonable range by applying an event-specific multiplier, 1.84523 for the high jump. So, concretely, the points earned for clearing a height of x centimeters in the high jump, as long as x is greater than 75, is 1.84523 times $(x - 75)$ raised to the 1.348th power. That's right, the scoring system for *each event* in the heptathlon is a formula that is more complicated than Einstein's $E = mc^2$.

This choice of scoring system makes the heptathlon select for relatively well-rounded athletes, which is arguably the goal of the sport. Competitors pay close attention to the scoring system when deciding how to allocate their training time. Is it better to shave a tenth of a second off your current time on the hurdles or the 1,000-meter run? The scoring function relates both potential outcomes to the bottom line.

That's how it works with people. How do reinforcement learning algorithms use objectives to improve *their* performance? I recently ran across a concrete example that I think is really informative: golf. (I know, I'm leaning on sports analogies a lot. But trust me, this one is good.) Let's say you are working on your golf game. It can take five or six strokes (hits on the ball with a golf club) to go from the starting position at the tee to getting the ball in the hole ("holing out"). Fewer strokes is better than more strokes, so we can tell that completing a hole in three strokes means you did something good and completing the same hole in eleven strokes means something—probably a bunch of things—went wrong. But this overall score doesn't really tell you where to focus your practice time.

Columbia business professor Mark Broadie pioneered a scheme for taking the overall performance on a hole and breaking it down into evaluations of the individual strokes a player takes. It's called "strokes gained," and goes like this. A team of people who are big fans of both golf and math analyze each hole on each major golf course. They work out, for each location on the course, how many strokes, on average, the pros take to hole out starting from that location. For example, they figure out that, from a spot on the fairway 240 yards from the hole, players average 3.44 strokes to hole out. From a spot 180 yards from the hole but in the rough, players average 3.29 strokes to hole out. The team maps out the whole area so that, no matter where the ball is, they know the "strokes to go."

Let's say you are playing a hole and you find yourself 220 yards out and in the rough. You line up your shot and hit the ball and it lands on the fairway, 30 yards from the hole. Sounds like you did great! Let's check. The strokes-to-go for the initial location was 3.51. The strokes-to-go after the hit was 2.50. So, in one stroke, you made 1.01 strokes of progress toward holing out. That seems, well, pretty average. (I mean, it's average for pros, so if you aren't a pro, it sounds pretty awesome.) In general, we can measure performance for a stroke by taking the strokes to go before the shot minus the strokes to go after the shot, minus one (for the stroke actually taken). If the number is positive, it means you are beating the average—nice work! If it's negative, it means you are losing ground compared to other golfers, and maybe that's not so great. Basically, the existence of the strokes-to-go map lets us translate the overall objective to one that applies to individual actions so that we can do credit/blame assignment and learn to improve.

In reinforcement learning, this difference in score across time is known as the temporal difference and it plays a role in all major approaches to learning good behavior. The inspiration for creating learning algorithms that can leverage experience to adapt behavior comes from the study of animal learning in psychology. Folks like Pavlov discovered that animals, including people, change their behavior in response to certain stimuli. Some stimuli, like a sip of juice, result in strengthening or "reinforcing" the behavior that led to it. I taught Sadie to jump through a hoop by providing her bits of chicken for approaching, stepping through, and eventually jumping through the hoop. The chicken served as a reinforcer because when the dog did the right thing, she got chicken, and that made it more likely that she would repeat that behavior in similar situations in the future.

Computer scientists Rich Sutton and Andy Barto found that idea quite powerful and set about figuring out how to make computer programs that could learn behavior from experience. They adopted the term "reinforcement learning" from the animal learning researchers, but they meant something different by it. Instead of focusing on stimuli that strengthen behaviors, they framed the problem as one of reward maximization—stimuli like chicken or praise and scratches behind the ear have reward values and the animal/algorithm is trying to get as much reward as possible. Basically, computer scientists stole the "reinforcement learning" term but misused it in the process. That's kind of embarrassing for us, but we've had the last laugh. Neuroscientists have discovered that the temporal difference concept that is so central to computational reinforcement learning is implemented in the brains of animals through the neurotransmitter dopamine. Specifically, dopamine tracks the temporal difference: More dopamine signals that the rewards were greater than expected (strokes gained!) and less dopamine signals that the rewards received were smaller than expected (strokes lost . . .). Maybe reinforcement learning was never about reinforcement at all! Nowadays it's pretty common to hear a psychology researcher talking about reinforcement learning and mean it in the computer science sense of the word. Knowing that my field has helped brain scientists better understand how behavior learning works . . . well, that really gets my dopamine flowing.

THE BENEFITS AND COSTS OF LEARNING FROM COSTS AND BENEFITS

Using a reinforcement learning approach for conveying loopy tasks has some important properties. First, defining the task by a reward function or evaluation score allows for more creativity and open-endedness in the resulting behavior—the computer can surprise us with its ingenuity. Second, the character of the solutions is shaped by the choice of evaluation function—the algorithm responds to incentives. Third, it's still quite hard to select an evaluation function that encourages the behavior you might want.

Here's an example from my robotics research of a reinforcement learner finding a creative solution. There has been some success using reinforcement learning to design robot behaviors for low-level skills—balancing a stick, hitting a ball with a bat—perhaps not things you *need* a robot to do (unless you happen to be starting a robot baseball league, which sounds AWESOME), but things that might help researchers figure out if they are on track to solve related but more practically relevant problems, such as helping to back up a tractor trailer or creating a robot that can reach for a doorknob. My students and I built a robot dog escape room to study high-level decision-making. Our robot dog was placed in a wooden pen. On the wall of the pen there was a button that, when pressed, opened a sliding door through which the robot could exit and earn ten reward points. One point was deducted for every second the robot was still inside the box. The pen was closed on top, making it too dark inside to see the button. Fortunately, a second, lit button could be pressed to turn on the lights. We wanted the robot to learn to turn around in the box until it could see the lit button, approach the button and press it with its snout to turn on the lights, turn around to find the now visible door button, approach that button, and press it to escape. The length of the required sequence of decisions would stress test our reinforcement learning algorithm and demonstrate that it could learn effectively in a real environment.

Since we penalized the robot for the time it was trapped in the box, the robot had an incentive to find a shortcut if one existed. And one did. Because the buttons were permanently mounted on the walls of the pen, their relative positions never changed. The robot learned to use the lit button to orient itself and, while looking at it, walk backward into the door button, pressing it with its robot dog rear end. We didn't know it was even possible for the robot to put its butt on the button. We had designed a specific snout-based movement it could use for pushing the buttons. But the robot had ample time to experiment in the pen. It kind of backed into this (objectively better) alternative solution by accident.

Reinforcement learning has been tested in a number of domains where the learner has a chance to learn something new or at least useful: Elevator control (figure out which elevator to send to pick up which passenger to try to minimize overall waiting time for everyone), circuit design (figure out how to place all of the relevant components onto a circuit board to minimize manufacturing cost), and thermostat control (keep the room temperature in the target range while minimizing energy costs). But the applications that have gotten the most attention are those in which reinforcement agents learn to play (and win) games that people play.

Games like chess, backgammon, and go hold a special fascination for people because they are often associated with world-class feats of mental discipline. Chess grand masters, even prodigies, train for years to become the best players they can be. Being beaten by one produces a strange feeling of helplessness. Even though you have access to the same pieces and the same rules, somehow your opponent handily outmaneuvered you and, once you see the corner they painted you into, it is too late to do anything about it.

Board games are also very naturally formulated as reinforcement learning problems. The actions are the legal moves of the game, the state features constitute the board itself, and the objective . . . well, it's to win. My very first reinforcement learning project was something I worked on during winter break in college in 1985. I was interested in programs that could learn, and I thought that getting a program to learn to play tic-tac-toe was just about the right level of difficulty for something to try—it's a complex enough game to keep kids interested, but it's still a pretty tiny game, all things considered. (The Wikipedia page for the game provides the optimal strategy in the form of a single-picture nested diagram. I highly recommend taking a look at it just for the sheer elegance of the design.) I stayed up until the wee hours coding my tic-tac-toe game. Then I started the learning algorithm playing against itself, and went to bed. I fashioned myself as the protagonist of the movie *WarGames* teaching the AI computer WOPR that "the only way to win is not to play." I awoke the next day to discover my learning algorithm did indeed discover that guaranteeing a win in tic-tac-toe is impossible, and, in response, just played randomly, moving into the first empty square of the board no matter what the rest of the board looked like. It played as though it just didn't care anymore. Sometimes it won, sometimes it tied, sometimes it lost. What difference did it make, really? Well, that was definitely not what I was going for. It took another late night of programming before I figured out the trick: if it can't force a win, at least get it to avoid a loss! The next morning, I awoke to find my program had taught itself to become a tic-tac-toe master.

The coolest thing that my algorithm discovered was how to make a "fork"—a board position where the computer had two ways to win. No matter which way the opponent blocked, the computer could always play the other route for a win. It didn't learn the concept of a fork in any generality. It just tried to play to reach any of a set of specific board positions that had this special property. (A decade later in my doctoral dissertation I was able to prove that this approach to learning by playing against yourself achieves what is known as minimax-optimal play—it guarantees a win when it can and does its best to avoid a loss if it can't.)

Just a few years later, IBM researcher Gerry Tesauro was breaking new ground in machine learning. He had designed a program that learned to play backgammon by imitating experts. His Neurogammon program used a clever application of the kind of neural network training described in chapter 5. Specifically, it was trained on expert move decisions (a board and the expert's choice of move from that board). It was configured to map board states to evaluations ("How good is this board?") so that the boards resulting from the expert's moves scored higher than those resulting from the other possible choices. The result is that it constructed an analog of the strokes-to-go function from the golf example. After learning was complete, it could play new games by choosing actions that its evaluation function scored higher than alternatives. And it worked quite well, defeating the best hand-programmed backgammon programs of its day. But, trained on data from human decisions, it was still a pale imitation of expert play.

Gerry decided to see whether this newfangled temporal difference idea might eliminate the dependence on human data and maybe even improve overall play. He created TD-gammon, a neural network that learned an evaluation function by self-play through the *temporal difference* approach. It was a gutsy move. At that time, TD learning had only been applied to artificial prediction problems akin to "If there's five in the bed and the little one says roll over, how likely is it for one to fall out?" Nevertheless, Tesauro got it to work. TD-gammon not only surpassed Neurogammon's level of play, but sparring matches with top human players suggested it was among the best backgammon players of all time. It spurred an explosion of interest in reinforcement learning and it also taught human competitors some better backgammon strategies that improved their play. This sequence of events was repeated again about twenty years later with the game of go and an intrepid team at DeepMind led by David Silver.

One of the wonderful things about conveying a task to a machine through reinforcement learning is the notion that it can learn truly new strategies that no computer or even person had previously identified. That concept is really inspiring. There are lots of problems that humankind could use some help solving. If reinforcement learning programs can outplay humanity's greatest players in board games, maybe they can also do better than we can at, say, stopping global warming.

There's a dangerous logical fallacy at work here. Just because a machine can solve a given hard problem doesn't mean it can solve every hard problem or even that every hard problem is solvable. But even if such a thing were true, there's a deeper issue that needs to be considered. Board games are very well circumscribed. We know the actions and the relevant state information, and the objective can be written down formally. It is quite different from real life where all these components are subject to debate. Let's consider the tic-tac-toe example again. If we keep the objective fixed as "get three in a row," but we expand the action set from "moves on the tic-tac-toe board" to "unrestricted access to the internet," the game changes dramatically. In principle, such a tic-tac-toe program could figure out whom it was playing against, hack into their bank account, and blackmail them into throwing the game. I'm not aware of any learning algorithms behaving this way, but history is filled with examples of people going outside the rules of games to improve their chances of winning. And a reinforcement learner in one of my projects did exploit a hole in the university authentication system to get itself online "illegally." My point is that a move to open-ended decision-making makes the reinforcement learning approach harder and potentially much more risky. Current learning algorithms have free rein to use any available action to try to maximize reward, making it challenging to walk the line between giving the algorithm enough flexibility to do something interesting and giving the algorithm enough flexibility to undermine the actual goal it is asked to achieve.

MY ENDLESS LOOPS

Sequences (chapter 3) and loops (this chapter) are kind of the yin and yang of computation. Sequences are the yang (阳)—pressing forward, unrelenting. Loops are the yin (阴)—receptive, flexible. Just as with yin and yang, each is contained within the other.

A loop can be unrolled into a sequence with the steps from one iteration to the next explicitly laid out. A sequence can be expressed as a loop, where each step in the sequence becomes its own iteration of the loop: On iteration *i* of the loop, do step *i* of the sequence.

As a concrete example, let's consider Shirley Ellis's 1964 "Name Game" song. Combining command sequencing (chapter 3), the variable "the name" (chapter 5), and a few conditionals (chapter 4), a verse of the song consists of twelve steps:

```
1 say the name
2 say the name
3 say "bo"
4 say "b" if the first sound of the name isn't "b"
5 say the name without the first sound
6 say "bo-na-na fanna fo"
7 say "f" if the first sound of the name isn't "f"
```

Figure 6.2 The black-and-white swirl picture connotes the interplay between yin and yang. *Source:* Image by DonkeyHotey, Wikimedia Commons, CC-BY-2.0 license https://creativecommons.org/licenses/by/2.0.

```
 8  say the name without the first sound
 9  say "fee fi mo"
10  say "m" if the first sound of the name isn't "m"
11  say the name without the first sound
12  say the name
```

(My mom used to sing this song for my siblings and me when we were growing up. We'd call out names and she'd sing the verse. Fun! Interactive! As we got older, we thought it would be funny to have her do "Chuck." She got to steps 7 and 8 and realized we had probably outgrown this activity.)

That's the song as a command sequence. But we can also render it as a loop:

```
for i from 1 to 12:
    if i equals 7 AND the first sound of name isn't "f": say "f"
    if i equals 9: say "fee fi mo"
    if i equals 3: say "bo"
    if i equals 10: say "m" if the first sound of name isn't "m"
    if i equals 4: say "b" if the first sound of name isn't "b"
    if i equals 5 OR i equals 8 OR i equals 11:
        say name without the first sound
    if i equals 1 OR i equals 2 OR i equals 12: say name
    if i equals 6: say "bo-na-na fanna fo"
```

The idea is that, each time around the loop, we check which step we're on (the state of the sequence) and then say the appropriate thing for that step. I scrambled around the order of the tests to emphasize that the order doesn't matter. In this example, turning the sequence into a loop increased the overall length by 50 percent: even though we saved a few words because "say the name" and "say the name without the sound" only needed to be said once instead of three times, all the step testing ("if *i* equals 3," for example) adds a lot more words.

Nonetheless, there are still some pretty good reasons to use loops to express some instructions that could have been written as a sequence. For example, we could tell a robot to build a five-block tower for us with the sequence stack-stack-stack-stack-stack. Or we could say "while there are fewer than five blocks in the tower, stack." This second option will continue to work even if someone helpfully adds a block to the tower while the robot is working or if the robot knocks the tower over by accident during the stacking process.

The idea of thinking of sequences as actually being loops that are checking and perhaps updating state isn't just a cute trick. It's absolutely central to the functioning of modern computers. The hardware of the computer itself is just executing a simple loop, billions of times a second. The loop is:

```
do forever:
    fetch the next instruction
    do the instruction
```

Everything the computer does, from sending emails to editing images, to executing database searches, to running fancy new learning algorithms, goes through that narrow little loopy bottleneck.

Loops are the final computational unit needed to cover all of what is necessary to specify arbitrary tasks. But there is one last concept that helps make specification clearer and more cognitively wranglable: functions. We'll see that they also give us one more way to loop.

7

DEFINING FUNCTIONS
CALL ME BY YOUR NAME

In the movie *Repo Man*, a character expresses the idea that once you recognize something as a concept—stores selling plates of shrimp, for example—suddenly you start seeing that concept everywhere. And indeed, an ad for a plate of shrimp is seen in a restaurant window in a later scene. I thought that was just something they made up for the movie, but I now know this phenomenon as a version of the *Baader-Meinhof effect* and see it all over the place: once you give something a name, it becomes "a thing," kind of like how the phrase "a thing" became a thing around 2009.

When we convey tasks, especially complex ones, giving a name to a particular behavior provides a number of important benefits. I focus on three of them in this chapter—*providing a shorthand, managing complexity,* and *supporting recursion.*

Chapter 3 introduced macros as a convenient way to package a sequence of commands into a single unit for later use. User-defined functions, sometimes called procedures, provide the same capability for more general behaviors built out of any of the concepts we've covered— conditionals, loops, and so on. Just as car radios used to come with pro- grammable buttons ("I'll use button 1 for a top-40 station, button 2 for news, button 3 for country . . ."), you can associate whatever behaviors you want with whatever names you want to give them. Later, when you want to initiate that behavior, you don't need the details of how to do it anymore—just the name. It becomes shorthand for a more complex action you wish to convey.

When we talked in chapter 5 about cooking a plate of shrimp in a hibachi restaurant, I emphasized how variables provide a kind of shortcut for referencing the results of a sequence of steps, so that the steps could be taken once and the value used multiple times. Functions are like that as well, except they refer to the steps themselves and not just the values produced by those steps. They let you reuse an entire complicated activity—including, for example, the loops we discussed in chapter 6—just by invoking their name. A function, once defined, can be treated like a command of the kind we discussed in chapter 3—something that is available for use whenever it's needed. It can even be used in the definition of *other* functions.

Let's look at an example of how defining functions—naming an activity—can be used to tame the complexity of a difficult task.

ON SCRIPT

In chapter 1, I complained about the fact that software today is less modular, recombinable, and programmatically customizable than in the old days, when it was better set up for you to tell it what to do. I stand by that. But to be fair, there are little glimmers of light shining through in some modern software. Apple Script and Visual Basic Script let you send commands to programs running on your Apple or Windows computers. As a result, you can automate the effect of clicking on things with your mouse to produce behaviors that would otherwise require your direct attention. Adobe, maker of Photoshop, Illustrator, and other popular "creative" programs, supports this kind of scripting across its products so its users can customize their own workflows and, in Adobe's language, create an assistant that "happily does the mind-numbing tasks," such as using Adobe software. Adobe also lets you express commands in the language JavaScript, which works across different computer operating systems.

Other software supports scripting as well. As I mentioned before, I do a lot of my work in Emacs (ahhh, Emacs . . .), which is scriptable, but also Google Apps—Docs, Slides, Sheets, Forms, and so on. I recommend these applications because they are technically excellent, well supported, and free. They also are scriptable. (Note, though, that Google's offerings are "free as in beer" to you, but to Google they are "free as in free to track your every move and try to sell you things." Caveat, non-emptor!)

Code to Joy: Why Everyone Should Learn a Little Programming

1. Telling Computers What To Do: I've Got You (6.5k words)
2. The What of Programming: Allow Me To Introduce Myself (4.6k words)
3. Sequencing Commands: I'll Take Your Order Now (10.5k words)
4. Splitting on Conditionals: If That's What You Want! (8.9k words)
5. Storing in Variables: It's What I Stand For (9.3k words)
6. Consolidating into Loops: You Don't Have To Tell Me Twice (11.1k words)
7. Defining Functions: Call Me By Your Name (8.4k words)
8. Combining Code and Data: Sure, But Can You Give Me An Example? (9.7k words)
9. Programmable World: As You Wish (4.3k words)

Acknowledgments: Thanks! (0.3k words)

Figure 7.1 Throughout the writing process, the table of contents document for the book maintained word counts so that the author could keep track of progress.

Google Apps Script is the language Google provides for creating new behavior in Google Apps. I've used it to help my teaching assistants organize grading information in Google Sheets and to keep track of the length of lectures I've written in Google Docs. I decided that updating the word counts for the chapters of this book could be a nice demonstration of the kind of functions you can write in Google Apps Script.

I drafted each chapter of the book in its own Google Doc and created a separate Google Doc as a table of contents with links to each of the chapters. This table of contents is just a plain old Google Doc except it lists the title of each chapter, contains a link to each chapter, and notes the length of each chapter (in words).

MIT Press gave me a maximum word count for the book, which translates to about 8,000 to 10,000 words per chapter. For a while, I was running a word count function on each chapter, then typing the result into the table of contents document so they'd all be in one place together. It didn't take long for this sequence of actions to become tedious. In fact, sometimes I'd make an edit to one of the chapters, forget which chapter it was, and then have to redo the counts on all of them to make sure they were up to date. No one told me that the hardest part of writing would be the math!

Extending the behavior of Google Docs to automatically keep the word counts up to date in my table of contents document would be a pretty handy bit of automation for me, so let's do that. Along the way, we'll get a demonstration of why defining functions is so useful, and we'll also see some old friends we met in previous chapters.

Here's the plan for automating the process for keeping my table of contents document up to date with accurate page counts for each chapter:

1. Scan the table of contents, finding links to the chapter documents.
2. Use the links to get the text of each chapter.
3. Count the words in each chapter.
4. Update the word count in the table of contents document.

We'll write functions for each of these because we're going to use a loop—chapter 6—to visit each chapter and perform the same actions, conveniently packaged into functions that can be invoked by their names.

But before we dive into the example, it's important to have some understanding of the structure of a computerized document. The text in a Google Doc is arranged into a kind of tree structure . . . a bit like a pharmacy. In a pharmacy, you have a main sales floor, which is divided into sections like "pain relievers" and "eye care." Each of those section has a set of shelves. Each shelf has the products arranged by manufacturers in groups. Finally, there are the products themselves—contact lens solutions, laxatives, and so on. The rationale for this scheme is that grouping things makes specific items easier to find than if everything were just in one big pile. You want Visine? It's better to go hunting in "eye care" than in "cold remedies."

Similarly, all the text in a word-processing document is organized to make it easy for programs—and, through the programs, people—to modify the document. In a typical Google Doc, the text is in the *body*. The *body* very likely contains a bunch of *paragraphs*, *lists* for one-dimensional organizations of text, and *tables* for two-dimensional organizations of text. A table is broken down further into *rows*, each of which has *cells*. A cell is like a tiny *body* in that it can also include paragraphs, lists, and even another table. At the very bottom of this structure is *text*. Text can have properties, such as font, size, and color, but in Google Docs, it's unable to nest anything else within it.

With that structure in mind, let's write a function that updates my table of contents document with the latest word counts. We'll start from the end and work our way up and out. So, what's the last thing we want to do in this process? Replace the word count in the appropriate line of the table of contents document. That assumes that we have functions that have already looked at the chapters and counted the words. (We'll get there shortly.)

At one point, my table of contents had a line of text that read

Chapter 6: Defining Functions: Call me by your name (9.0k words)

But I made some edits, and the new word count became 9.4k. Replacing the old count only takes a function that's two lines long using a Google Apps Script. Unfortunately, the syntax for Google Apps Script isn't particularly friendly. It uses a style of writing code based on JavaScript, which is heavily influenced by a lineage that began with a language called C in the early 1970s. Wikipedia lists about seventy-five computer languages in this family, which was devised for computing professionals. So, instead of presenting Google Apps Script code, I'll stick to the easier-to-read style I've been using throughout the book. It's actually inspired by Python, which is a very popular language these days, with hints of the user-friendly language Scratch thrown in:

```
define the function edit count count in text:
    delete pattern '(xk words)' in text
    append '(xk words)' where x is round(count/1000) to text
```

The very first line creates a new function and names it _edit count_. That's the name I'll invoke it with later on. I can name the function anything I want, but I'll regret it later if I use names that don't hint at what the function actually does. ("Wait, why are all these functions named for different brands of toothpaste? What does the _Sensodyne_ function do, again?") In this case, the function rewrites the text element it is given to display a new count. So _edit count_ seems like a reasonable name for it.

Following the name is information it needs to do its job: two variables, _count_ and _text_. These are the _parameters_ of the function. If you want me to edit text to change the count, I need to know what text to edit (the _text_ variable) and what the new count should be (the _count_ variable). I ought to be able to remember what they stand for. This notion of parameters is really important, and it builds on two ideas introduced in chapter 5. First, when the function is run, the initial input for these variables is set by how _edit count_ is invoked. If we say "_edit count_ 4557 _in table of contents_," _count_ takes on the value of 4557 and _text_ takes on the value _table of contents_. That gets things started. The second idea that's used here is that of _scope_. While the instructions in the function are running, the variables _count_ and _text_ refer to these new values and not anything else that might have been in use up to that point. The function creates a new scope for these variables so we don't have to worry about them interfering with any other instructions that are going on. Getting a complete feel for how this works is tricky and beyond the scope of this book. It's one of the subtle distinctions between different programming languages. But for the purposes of this example, the idea is relatively straightforward—in the function, _count_ and _text_ refer to these new variables and whatever values they are given when the function is run.

The next two lines of the function constitute the code that should run when anyone asks to do "_edit count_." The code is a command sequence, as we discussed in chapter 3. The first command in the sequence is **delete pattern**, which deletes any matches to the pattern in the given text.

The _edit count_ function is outsourcing the actual editing of the text to another function. The **delete pattern** function is built into Google Apps Script, which is handy because coding with functions encourages this kind of delegating.

Executing this command on our example text, it notices that the "(9.0k words)" matches its pattern and deletes it, leaving the text as:

Chapter 6: Defining Functions: Call me by your name

The second command in the _edit count_ function assembles the text of the new word count message and attaches it at the right spot. The message is the given _count_, divided by 1,000 and rounded, inserted into "(_x_k words)" where the _x_ is. (There's another handy use of variables!) The **append** command glues the message to the end of the text element, modifying the document so that it now reads

Chapter 6: Defining Functions: Call me by your name (9.4k words)

That's the desired behavior, and the way I wrote the function means the computer will be able to do the same kinds of edits to any text element and count that I give it. In a sense, I've taught the computer how to do a new thing, expanding its set of capabilities.

MAKE IT COUNT

It's all well and good to have a function that can paste a new word count into the table of contents, but where does that count come from? (Transylvania, of course, ah ah ahhh!) If I have to count the words myself, that would not be very helpful. Once again, we'll define a function to capture the needed behavior:

```
define the function get word count of url:
    set space to ' '
    set doc to open document url
    set contents to get body text of doc
    set words to split contents into a list at space
    set word count to length of words
    return word count
```

This function takes in the value of the parameter *url*, which will be the URL of a document containing the chapter that I need the word count for. (URLs are universal resource locators, what most people call a "web address." We'll dive into them a little later in the chapter.)

It then assigns values to a series of variables, one per line, similar to the sequence of variable assignments that make up the neural network in chapter 5. The first variable represents the character that separates words, which is a *space*.

The second variable is doc, which is a label for the chapter document itself. The open document function is provided by Google Apps Script; it turns the web address of the chapter (the one indicated by the *url* variable) into something that can be read and written by other commands.

With the document in hand (in variable?), we can get its contents using the Google Apps Script function get body text. The split command then turns that sequence of characters into a list of words as separated by spaces. I refer to that list using the variable *words*. The number of words in the list *words* is the word count of the document! So the next line runs a built-in function called <u>**length**</u> on the list of words and names the resulting value *word count*.

The value in the *word count* variable is the prize we sought. The final step is to deliver it to whoever it was that asked the <u>*get word count*</u> function to calculate this value. That's done in the last line using **return**. The **return** statement signals the end of an assembly line: We took in some raw material (the parameters), processed them into some intermediate forms (variables), and finally produced an end result (the value being returned). This particular function returns the length of the list of words it constructed, which is the word count of the document.

Now, where did <u>*get word count*</u> get the URL of the chapter it's counting the words of? It comes from the table of contents, but we need another function that scoops up the URL associated with a link in the text in the table of contents—if there is one. The function features a conditional, of the kind we examined in chapter 4. (Things worked out really nicely so that each function gives one of the prior chapters a curtain call.) Here's the new function:

```
define the function get link from text:
    set url to get link url from text
    if url is not null:
        return list consisting of (url, text)
    else:
        return empty list
```

The new function, <u>*get link*</u>, has one parameter, *text*, which is a text element. The function's task is to get the URL associated with the text element and return it as a list. The function creates a variable called *url* to hold any URL the built-in function <u>**get link url**</u> finds. If no URL is found, the *url* is given the value null, a value many computer systems use to denote a kind of nonvalue value like this.

The *if* statement compares the value of *url* to null. If they are not equal, the function knows that a URL was found. In that case, it returns a list containing the text element and its URL, paired up. On the other hand, if no URL is found, the function just returns an empty list. That completes the transformation from a text to a URL list.

The three functions I've written so far are examples of using functions as a *shorthand* for more complex combinations of other commands. The function *get link* (feat. conditionals) is presented with an element of text from the table of contents and pulls out a URL link to a chapter, if there is one. The function *get word count* (feat. variables) follows a URL link to work out how many words there in the chapter linked to that URL. The function *edit count* (feat. commands) is given a word count and an element of text in the table of contents and updates the text with the new word count. The three functions are just crying out to be daisy-chained together so that a text element has its URL extracted by *get link*, the words of the document at that URL are counted by *get word count*, and that count is inserted back into the text element by *edit count*. That's where I'm heading. But I'm going to break it down a little differently. I'm going to write a function *get links* (feat. functions) that will scour the table of contents document for links, compiling them into a to-do list, and a function *edit counts* (feat. loops) that goes through the URL to-do list, making each of the edits.

Let's look at the function **edit counts** first; it's simpler and it illustrates one of the other key function concepts: *managing complexity*. It starts with a compiled list of the URLs of the documents containing each of the chapters (to be assembled by *get links*), as well as the text elements in the table of contents that contain the chapter titles. That list will be the parameter *url list* to the new function:

```
define the function edit counts in url list:
    foreach (text, url) in url list:
        numwords = get word count of url
        edit count numwords in text
```

The function uses a *foreach* loop to run a short sequence of commands on each text/URL pair, which will correspond to the separate chapter documents. The first command takes the URL and sends it to the function *get word count*, which returns the number of words in the document with the given URL. The word count that comes back is assigned to a variable called *numwords*, which the function sends along to *edit count*, along with its associated text element, to make the necessary edits in the table of contents. That's the loop. It makes one edit for each of the chapters, after figuring out the number of words that chapter contains.

Let's take a moment to appreciate how (relatively) direct this last function is. Or, more precisely, to consider how messy and confusing it would have been to write *edit counts* without the get *get word count* and *edit count* functions. The complexity of each component is compartmentalized, which meshes nicely with how we wrap our minds around complicated ideas in general.

That's the wonder of functions. We went through the effort of defining *get word count* and *edit count* and now we get to just wave them around like magic wands and they do their thing for us. Think of it this way. Every time you write a function, you are standing on the shoulders of giants, building on top of any other available function. And you become a giant yourself, with your shoulders available for future function writers, perhaps even yourself, to benefit from your efforts when solving even more complex problems. What could be more friendly?

The way we use functions to express tasks strikes me as being a little like what you are doing right now: learning about how to tell machines what to do by leveraging your existing knowledge of telling people what to do, your experiences using computers, even the time and energy you spent learning to read. The beauty of functions is that every new one you write becomes part of a giant pyramid of task knowledge making more and more complex behavior easier and easier to express.

TALKING ABOUT TALKING ABOUT RECURSION

My new functions work together to turn a list of chapter URLs into edits to my table of contents document. But it remains to gather that list.

I could take a shortcut and look for the URLs in very specific places in the table of contents document, but it's kind of cool to look at how to handle the task in greater generality. Recall that word-processing documents have a nested structure, with paragraphs, lists, tables, and so forth. The tables can themselves contain lists and, yes, even other tables. Table-ception strikes again! In principle, there's no limit to the number of levels we can nest our tables and there could be chapter links inside tables that need to be gathered. After all, since I'm searching the table of contents, I should be sure to include the contents of the tables. So, to find URLs in a document, wherever they might be lurking, no matter how nested they might be, I'm going to use recursion.

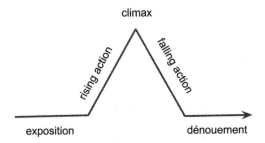

climax

rising action

falling action

exposition   dénouement

Figure 7.2   Freytag's pyramid diagrams a standard structure of stories.

Recursion? To curse again? No, although it has that effect on many people, including some top-notch computer science students. Let's take an excursion into recursion . . .

We saw in the `edit counts` example that the definition of a function can make use of a previously defined function. In this context, recursion takes the form of writing a function that makes use of *itself*, a function that hasn't even been written yet. That sounds like it should be illegal or, at the very least unwise. Wouldn't it produce a never-ending loop that just drills itself further and further into the ground until it comes out the other side and then pops up and just keeps going and continues forever? Not necessarily. You may not be familiar with a function that has a copy of itself inside it, but you're familiar with stories that have stories inside them. Let's look a little deeper.

After all, storytelling is ubiquitous in human societies. Some evolutionary psychologists even see it as a key cognitive capacity that lets us work together and learn from each other and build societies. At least, that's one story I've heard. A standard way of diagramming story structure is known as Freytag's pyramid, which consists of four lines that read left to right.

First is a horizontal line representing the exposition, the part of the story that introduces the characters and setting. The line then bends upward to reflect the rising action in the story in which tensions or complications build. The endpoint of that line segment is the climax, which is a turning point for the main character, followed by a downward-sloped line representing the falling action as the conflict resolves. Finally, a new horizontal line appears corresponding to the resolution of the conflict or dénouement, more Frenchly.

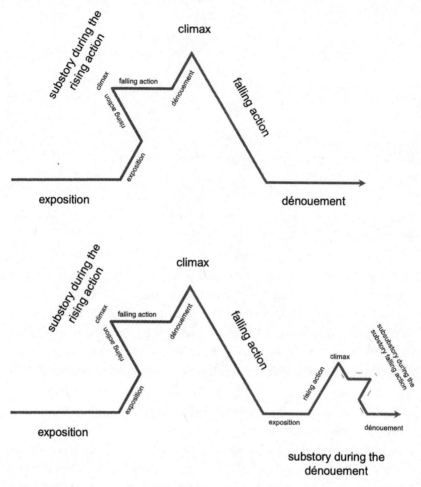

Figure 7.3 Top: A story diagram can show a story that has a substory in it. Bottom: A story diagram can show a story that has a substory in one of its substories.

Here's the wrinkle: Any one of these lines can be its own story, also a Freytag pyramid. A story can be made up of other stories. Recursion!

We can see this idea in "Brandy (You're a Fine Girl)," a song that was a chart topper in late August 1972 that tells a little story. Feel free to sing along, if you know the words. In the exposition, we meet a young woman who works in a port town. Brandy serves, and is apparently named for, alcohol. In the rising action, patrons at the bar express their interest in marrying Brandy. However, her heart belongs to another, a man whose name is on the locket she wears. In the climax, we learn her affections are unrequited as he is unable to give up his chosen profession as a sailor. In the resolution, Brandy walks home after the bar closes, struggling to accept her fate. It's a poignant tale, told in three minutes and E major.

What I find so interesting about this song is how the story itself is told.

The action starts with her serving drinks, but we quickly go into a flashback of Brandy's time with the man she loved. There's a story inside the story. And, interestingly, in that story, we learn the man would entertain Brandy by telling her stories of his adventures, describing what it's like to be out on a stormy ocean. Before we hit the climax and the man declares that he chooses the sea over Brandy, we're three stories deep. And here I am right now telling you the story of a young woman who is thinking back on a story of being with a man who regaled her with his stories of being aboard a ship.

It's crazy complicated, but it's also the most natural thing in the world. Listening to the song, you don't get tripped up by all the nesting, you just go along for the journey. And, as the song winds down, we leave the stories one by one—first the waves, then the man, then Brandy, who walks off into the darkness. Our minds are very comfortable with this kind of nesting. Stories are about what happens in our lives, and since stories happen in our lives, stories can have stories in them. Prohibiting stories from being recursive would be unnatural.

What about stories about prohibiting the act of prohibiting? An August 15, 2021 *New York Times* article covered a legal ruling concerning public health measures to combat COVID. The first paragraph read:

The governor of Texas can ban mask mandates, at least for now, after the State Supreme Court sided with the state on Sunday, granting a request for an emergency stay of an appellate court ruling that would have allowed schools to make face coverings mandatory.

I had a lot of trouble following the logic here, which my colleague Bertram Malle, who studies social norms, described as "recursive deontic logic in Texas!" ("Deontic" means having to do with laws or principles.) So let's break it down piece by piece. I find that it helps to put names to things. That's the theme of the chapter, after all.

Clay Jenkins, a Dallas County judge, made a rule that people in public schools, universities, and businesses in Dallas County would have to wear masks through the current phase of the pandemic. That is, *Jenkins was mandating masks.*

The governor of Texas, Greg Abbott, didn't like that, so he banned all mask mandates in the state. Normally, a governor wouldn't be able to make a rule like that, but his powers were expanded temporarily, if ironically, by virtue of the fact that the state was experiencing a public health emergency—the pandemic. *Abbott was banning mandating masks.*

Bonnie Lee Goldstein, a justice in Texas's District Court of Appeals, didn't think it made sense to use one's emergency powers obtained to combat the pandemic to ban a public health intervention intended to combat the pandemic. She issued a ruling that denied Abbott's ban. *Goldstein was voiding banning mandating masks.*

The state supreme court considered the situation and decided to hold off on the ruling from the court of appeals, leaving Abbott's ban in place, at least temporarily. Nathan Hecht is the chief justice of the Supreme Court of Texas, which issued the stay. *Hecht was staying voiding banning mandating masks.*

Did that translation help? Where does it leave things? If you were a kid in school in Texas, would you have to wear a mask? According to my translation, no. Hecht said that Goldstein couldn't say that Abbott couldn't say that Jenkins couldn't say that masks are required. That means Abbott *could* say that Jenkins couldn't say that masks are required. And he did. So masks aren't required. Nesting unraveled. (As an aside, the school superintendent for Dallas, Michael Hinojosa, said he planned to continue the mask mandate anyway because the ruling didn't apply to his situation. *Hinojosa was ignoring staying voiding banning mandating masks.* So, yeah, wear your mask a little longer, kid.)

In the months that followed, legal wrangling on this issue continued. But the ultimate outcome is not the message here. It's the fact that we can continue the nesting process indefinitely. That's the essence of recursion . . . functions like "blocking" can be applied to anything, even the blocking of blocking of blocking of something. It's hurdles all the way down.

DRAWING ON RECURSION

I use the idea of nesting things inside other things all the time when I make presentation slides. Drawing programs provide a set of shapes you can use, such as circles and rectangles and triangles. You can change their color, their size, and their orientation. You can arrange three large white circles, two small black circles, one orange triangle, and two skinny brown rectangles into the body, eyes, nose, and arms of a snowman shape. The drawing programs also provide a grouping operation—select all the snowman parts, click "group," and boom, the separate pieces of the snowman become a unit. That's very much like what happens when you define a function—a sequence of commands pops together into a new command. In the case of the snowman, you can treat it as if it were a single shape. It can be copied, resized, and rotated. You can even arrange a small working group of snowpeople and group *them* together to make a snow team. Once again, the snow team acts like a shape, with all the snow individuals in it resizing, reorienting, and reproducing as a unit.

By nesting these simple shapes into snowpeople into snow teams, it becomes easier to *manage the complexity* of what would otherwise be hundreds of individual circles, rectangles, and triangles. Defining functions gives you that same leverage to create and manage solutions to complex tasks.

The ability to use one function as part of another function just by using its name opens up an intriguing possibility. Can you use the name of a function as part of its *own* definition? In dictionaries, that's considered

Figure 7.4 Three snow teams consist of seventy-two geometric shapes.

a no-no. If you define "round" as "shaped like a disk" and "disk" as "a surface that is round," we'd call that a circular definition and it's something to be avoided. In computational terms, though, it's called *recursion*, and it can be incredibly useful.

Beginning students of computer science struggle with recursion, which they find to be one of the first truly mind-wrenching concepts they are asked to absorb. Partly, I think this view is legitimate. I can name more than one professional computer scientist who can recount trippy experiences in college while learning recursion—suddenly seeing sidewalks or conversations with friends as infinitely deep nestings of flagstones or quips, for example. On the other hand, there are plenty of people who had trippy experiences in college who were not learning about recursion. There's something about being sleep deprived, learning lots of new stuff in a short period of time, and maybe contact with psychoactive substances that can really mess with your perception of reality.

These days, though, I think of recursion as less a mystical property specific to computation and more a mystical property of everyday cognition. It is a very natural extension of nesting where you don't put any limit on how deep the nesting can go. A description can appear entirely embedded in another description, for example. With a nod to *Notting Hill*, you can say:

I'm just a girl.

You can also expand on what kind of girl you are:

I'm just a girl standing in front of a boy.

Now that the boy is in the picture, you can expand on what kind of boy he is:

I'm just a girl standing in front of a boy who is looking for love.

You can expand on why he is looking for love:

I'm just a girl standing in front of a boy who is looking for love because his wife left him.

And so on:

I'm just a girl standing in front of a boy who is looking for love because his wife left him for someone he claimed looked like a movie star, which seems ironic because he himself has been on the short list for "Sexiest Man Alive" for a quarter century.

The big idea is that a description of a person can include, nested inside it, a description of a person. Nesting elements inside the same type of element is intrinsic to how we use language and even how we perceive the physical world ("This TV comes with picture in a picture!"). When conveying desired behavior to a machine, recursive functions let us match our description of what we want to the natural way we see nesting out in the world.

NAMING NAMES

Defining functions is about taming complexity by giving names to the things you are doing and then using the names to stand in for the details of those activities later. But sometimes naming things is itself a complicated activity (see the discussion in chapter 1 about my children and their commonly canine names or the discussion in chapter 5 about disambiguating the various Michael Littmans), and nesting can help mitigate this complication.

Every page, video, image, and so on accessible on the web has a name by which it can be accessed. The name is its universal resource locator, or *URL*. For example, here's the URL for a page that explains URLs:

https://csrc.nist.gov/glossary/term/universal_resource_locator

A URL consists of three different pieces. The first piece, "https" in this example, provides the protocol by which the resource item can be retrieved. The most common is "http," which stands for hypertext transfer protocol; it tells the web browser on your computer to ask for the file in a particular document-exchange language. Getting that format wrong can lead to confusion. The protocol "https" is a version of http that is secure—hence the "s"—in the sense that data is exchanged in a secret code so that it is unreadable to any computers in between the one asking for the item and the one supplying it.

Skipping the second component temporarily, the third component in the URL "glossary/term/universal_resource_locator" is the pathway to a file named "universal_resource_locator." Just as you can specify folders (sometimes called directories) on your computer, people running websites can organize their items into a folder structure. In this case, the file we're looking for is in a folder named "term," which is itself nested in a folder called "glossary." Grouping items and naming the groups helps keep the 50,000 or so pages on that site organized and navigable.

Let's go back to the middle piece of information in the URL, the hostname. The hostname is the name of the computer that can supply the item. Of all the computers in all the organizations in all the World Wide Web, that's where you need to go to find this particular page. In this example, it's "csrc.nist.gov." Reading the name right to left, "gov" means the site is run by a government organization; "nist," the National Institute of Standards and Technology, is the name of the organization; and "csrc," the Computer Security Resource Center, is the group within NIST running the site. (Computer security! No wonder they are using https!)

Every hostname has this basic structure, providing more specific ownership or control as we go from right to left. Interestingly, the ordering is the opposite of what we use in the filename portion of the URL, where the categories become more specific as we move from left to right. There's a real change of direction at that slash between "gov" and "glossary."

Naming hostnames this way is a powerful idea that has helped make the internet more tractable despite the tens of billions of separate computers it connects. If you want to connect to a particular host, say the web server for St. Joseph's College in Queensland's Educational System in Australia, based on its name, your computer checks its records to see if it knows where to find www.sjc.qld.edu.au. It probably doesn't, in which case it goes down the line from the St. Joseph host sjc.qld.edu.au to the slightly better known Queensland host qld.edu.au. That computer must know where to find St. Joseph computer. But, if your computer doesn't know Queensland either, it moves on to the Educational System host edu.au or even the site for all of Australia, au. If your computer doesn't even know how to get to Australia, there's one last step. It contacts the so-called DNS root that everyone knows. The host is responsible for keeping track of all of the hosts with just one name: au, us, com, org, and so on. No matter what, it will always recognize *someone* it can ask for information that will point it in the right direction.

It was not always so easy to name specific computers on the network. Nowadays, this scheme for naming hosts makes it possible for you to compose an email to mlittman@cs.brown.edu, and your computer can work out where to send it to reach me (again, if you want to send me a birthday card). In the 1980s, when I first got online, sending me a message required you to spell out the entire path the message would take to reach me. If the cognitive scientist Jeffrey Elman, a significant figure in the second wave of neural networks and someone whose email address from that era I was able to find online, had wanted to send me an email from his lab in San Diego, he might have addressed it to sdcsvax!ucbvax! decvax!yale!mlittman. Following this address-slash-instructions, the message would go from Elman's computer to sdcsvax, a Digital Equipment Corporation (DEC) VAX 11/780 run by the UCSD Computer Science and Engineering department nearby. The message would next be forwarded to ucbvax, another DEC computer running at UC Berkeley some five hundred miles to the north. The ucbvax computer would pass the message along to one it was in regular contact with, decvax, another VAX machine, but this one running at DEC headquarters in Maynard, Massachusetts, roughly three thousand miles to the east. That computer acted as a kind of hub for the northeastern US and would send the message to Yale's computer, which would deposit it in my local digital mailbox.

The modern naming scheme, in which more and more precisely targeted information is prepended to the front of the address, is a kind of recursive structure—a hostname is the name of a computer, followed by a dot, followed by a hostname. It's also an interesting echo of the distinction made in chapter 6 between providing imperative step-by-step instructions as in the old address scheme and traditional coding versus the declarative just-tell-me-what-you-want-and-I'll-work-out-the-details approach in the modern internet address scheme and machine learning approaches.

With this context about recursion and how it is used in computing systems under our belts, we can get back to writing our recursive function.

RECURSE WORDS

To automate the updating of the word count of each chapter in this book, I'm writing a set of functions that seek out the URLs in my table of contents, read the chapters those URLs point to, count the words, and replace the current word count text with the new one. We've written functions for all of this except the one that gets the URLs out of the table of contents document.

Here's that new function:

```
define the function get links from element:
    set links to get link from element
    for each child in children of element:
        concatenate get links from child to links
    return links
```

This function, get links, is short, but there's a lot of interesting stuff packed in there. It is given an element of the table of contents document, such as a paragraph. It uses my get link function to find the link associated with the element, if there is one. Then it begins a loop. The loop zips through all the child elements of the current element looking for URLs. A child element in the case of a document is any of the structures we've noted, such as paragraphs, lists, and tables. Great! But, as we've also noted, child elements can have children that themselves contain children, and so forth. And for this function to work on documents of unknown complexity, it will have to deal with that unfortunate fact of life.

Give this function an element of a document—a child—and it will find any URLs in its text. But suppose the child has children that might have URLs? Well, we have a function that finds URLs in elements—this very function. So, if get links is examining a child that has children, all we have to do is call get links again from within itself until none of the children have any children that haven't been examined for URLs. Recursion! It reminds me a lot of the sequel to the book *The Cat in the Hat* in which the cat is able to solve a problem by breaking it up into smaller problems and then deploying smaller cats (which he keeps in his hat) to solve them. Each of those smaller cats has the option of doing the same—breaking its smaller problem into even smaller problems to be solved by even smaller cats. As long as, eventually, all problems are small enough to be solved directly, the procedure will end. In our document case, that just means that, at some point, there has to be a table with no other tables inside it.

Once the loop has handled all the children (the small cats have finished their jobs), the compiled list of URLs is returned as the answer (the original cat is done, too).

So now we have the set of functions we need to find the chapters, count their words, and update the table of contents. We just need to run it. Fortunately, the Google Docs menu has an entry for Tools, which has an option of opening the Script Editor, an editor that definitely is no Emacs but has the advantage of being well integrated into Google Docs; if I want to run all these functions on my Google Docs, this is definitely the place to do it from. So, in the Script Editor, I paste all the new functions, along with one last function:

```
define the function update toc:
    set urls to get links from current document
    edit counts in urls
```

This function runs *get links* on the current document (the table of contents), then runs *edit counts* on the resulting URLs. With this function in place, Google Docs lets me set a trigger to run it at the appropriate time, just as in the trigger-action programs of chapter 4:

- **if** document opened: *update toc*

With this rule in place, I get updated counts whenever I navigate to or refresh the table of contents document. Building on ideas from every chapter of the book, including, appropriately recursively, this one, I've coaxed the computer into providing me with some valuable information that will help streamline my writing process.

If that seemed like a lot, we can again turn to machine learning to automate some of the effort. Summoning Godzilla!

CREATURE DOUBLE FEATURE

Traditionally, machine learning systems are one-offs, trained with task-specific data to solve one problem. However, more data and more computation make for better solutions, which means it can be very costly to solve problems one by one this way. Although machine learning algorithms don't give names to functions, the idea of organizing computations into reusable units is very important in modern applications of machine learning. It shows up in three different ways: *convolutions*, *architecture reuse*, and *pretrained models*. I'll illustrate these ideas in the context of creating a function for recognizing two different characters from photographs.

Growing up in the Philadelphia area in the 1970s, I watched a TV show on Saturdays called *Creature Double Feature*. Each week they'd put on two monster movies, back to back. I was pretty afraid of watching them, actually. But they came on right after the Saturday morning cartoons and often would start before I had a chance to get up and change the channel. If I missed my window for turning off the TV, I'd just have to cower behind a chair and wait until the monster movies were over.

The star of many of these movies was Godzilla. I never got great at telling Godzilla and his enemies apart. Which one was Titanosaurus? Which one was Gigan? I didn't know. They both kind of looked like Godzilla to me. They all had snouts and stood on two legs. They were greenish grayish, and I think they all had some kind of fins on their backs. They shrieked and they shot fire or lasers out of their faces. Eek! I'm scared just thinking about it. Between the similar-looking monsters and the dubbed dialog, I couldn't figure out what was going on.

It might have been useful back then if I had had some kind of recognition aid that could help me tell the different monsters apart. Then I could have asked a computer to look at the monster on the screen (it was hard for me to see it clearly from between my fingers) and let me know which one it was.

Writing a program that can tell apart Godzilla and Gigan is an example of visual object recognition, a problem that has been of central interest in artificial intelligence since the creation of the field. In 1966, MIT researcher Seymour Papert, whom we met in chapter 5, advertised for students to participate in his "Summer Vision Project," an attempt to create a significant portion of a visual recognition system with a team of students. He described how the group would break down the problem and, by the end of August, build software that would be able to identify objects like cups and tools in pictures. After all, we know a ton about the physics of light and how objects get turned into images. How hard could it be to reverse the process and turn images back into objects?

Figure 7.5 Godzilla and Gigan often fight. Godzilla shoots lasers from his head, and Gigan has atomic breath. Or maybe it's the other way around. *Source:* Photo 12 / Alamy Stock Photo. Photographer Archives du 7e Art Collection.

Pretty hard, it turns out. The problem was not solved in two months. In fact, after about two decades of people trying to write programs that could identify objects in pictures, they basically gave up, switching from coding to machine learning approaches to the problem. Even so, it wasn't until deep learning came along another twenty years later that object recognition got good enough to be practically useful.

The deep learning revolution began in 2012 with the success of AlexNet, the bot heard round the world. AlexNet was able to accurately recognize and name any of a thousand different categories of items in photographs, including types of animals, plants, activities, materials, and human-made objects, slicing the error rate by about half compared to the next best system. AlexNet is a neural network with 62 million separately trainable weights, carefully calibrated to produce unprecedented performance. Alex Krizhevsky, the leader of the effort, designed the pattern of connection in his neural network and trained it using the machine learning recipe discussed in prior chapters on a collection of approximately one million images—about one thousand examples for each of the categories it could recognize.

The first way that function-like reuse comes into play in AlexNet is through the notion of *convolutions*. Convolutions are little subnetworks that look for distinctive features in images, such as edges and gradients of colors, no matter where they appear. For example, to recognize a Siberian Husky, category 250 on the list, it's useful to pick out the distinctive blue eyes. Detecting the small round blue area with a black center is an operation that neural networks can learn to do very well. But, depending on how the dog is standing in the image, the eye could be anywhere from the center to the top right to the bottom left. Since each unit of a neural network is attached to some particular location in the image, there's no way to say "detect this pattern no matter where it appears in the picture." The network has to learn what the Husky eye looks like separately at each place in the image an eye can be.

So, it would be better if there were an eye-detector *function* that could be learned once and then applied at all image locations to see where the eyes are. That's precisely what convolutions do. Five of the layers of AlexNet consist of collections of automatically trained visual feature recognizers—functions—that it puts to use everywhere in the image. No one has to decide which recognizers pick up on which parts—the learning process works out all of that automatically. But arranging the neural units into functions this way saves on training data because the critical calculations need to be learned only *once*. That makes it possible to produce more robust and flexible object recognizers that continue to produce correct answers for items in photographs with varied lighting, poses, and contexts. As a result, Alex was able to train a remarkably powerful visual recognition system with only a thousand examples of each category instead of the millions or more that would be needed for a network without convolutional layers—learn once, use everywhere.

Second, visual recognition functions are reusable in other machine learning projects. For example, as soon as AlexNet was announced, it spawned variations. Researchers considered modifications of Alex's twelve-tiered arrangement of neurons, evaluating networks with more layers, fewer layers, bigger sets of convolutional patterns, different sizes of recognizers, and so on. Usefully, network structures, often called *archi-tectures*, that worked well for one specific collection of categories and data would work well for other visual recognition tasks. That means, if I wanted to train a network to tell Gigan and Godzilla apart, I could just take an existing architecture that has proven useful in the past, an AlexNet variant called VGG-16, say, from the Visual Geometry Group at Oxford University. It's not a complete function for solving my prob-lem, but it's essentially a function template that can get filled in through extensive training. It's worth a try.

Unfortunately, I was only able to find a dozen or so images of Godzilla and Gigan online. Even a carefully selected architecture with just the right set of convolutional layers can't be trained with just a dozen exam-ples. We're going to need to leverage the third way that function reuse is applied in deep learning, with pretrained models. Here's how it works. Let professionals train VGG-16 on millions of images until it can reli-ably recognize one thousand or so categories of objects. The profession-als do the work gathering the training data and spending the computer resources needed to train the neural network, perhaps millions of dollars if you include the search for the right architecture. They then give you a copy of this pretrained neural network. You say, "Um, thanks, that's very generous, but you forgot to include Godzilla and Gigan in your list of categories. Hello?," but they've already walked away and are working on some other problem, like speech recognition.

Somewhat amazingly, the pretrained network can still be quite useful. Since the network was trained to recognize so many different categories, it already knows about many visual features that are relevant to identify-ing Godzilla. It likely learned a brown eye detector for spotting St. Ber-nards (category 247), a fin detector for picking out tiger sharks (category 3), and what it looks like to stand tall and save the day by radiating heat and smoke, so that it can recognize espresso makers (category 550). As an experiment, I pulled five images each of Gozilla and Gigan off YouTube and called them my exemplars for the two creatures. Then I grabbed five images each of Gozilla and Gigan off YouTube and pretended that I didn't know which were which. I ran them all through VGG-16 to see which features were recognized and then matched each unknown image to its closest exemplar. Ten out ten times, the unknown image was matched to

its right exemplar—the pretrained version of VGG-16 correctly categorized all of my monster images.

Through convolutions, architecture reuse, and pretrained models,
machine learning approaches are using the function concept to turn data
into useful behavior that would be challenging if not impossible to produce any other way. In the next chapter, we'll look at how to make it
even easier for people to reliably convey tasks to machines; specifically,
we'll see how the complementary strengths of data and code can work
together in harmony to create something great, not unlike the epic part-
nership of plates and shrimp.

8

COMBINING CODE AND DATA
SURE, BUT CAN YOU GIVE ME AN EXAMPLE?

In high school algebra class, I remember being taught two different ways of defining sets of numbers: You can list out the elements or you can provide a general rule. So odd numbers might be written by examples like $\{1, 3, 5, \ldots\}$ or they could be expressed formulaically using a rule, as in $\{2k - 1 \mid k \in Z+\}$. The rule method has its shortcomings. Like, what the heck does that even say? Well, if $Z+$ represents the positive integers, 1, 2, 3, 4, and so on, we're taking each one of them, represented by k, doubling them to get 2, 4, 6, 8, and so on and then subtracting one to get 1, 3, 5, 7, and so on. I mean, it works, but it feels a little cryptic. Even odd. For working with a bunch of numbers known as the naturals, it seems pretty unnatural.

The example method seems much clearer here, but it too runs into problems. For one thing, there can be ambiguity. I say $\{1, 3, 5, \ldots\}$, but couldn't that also mean only the odd numbers that appear in the Fibonacci sequence: $\{1, 3, 5, 13, 55, 89, \ldots\}$? Or, for that matter, maybe it's the set of *all* numbers and I just gave a few examples that happen to be odd? Or what if I gave you a set like $\{1, 4, 6, \ldots\}$ and you don't even have a guess right away of what else is in it?

Beyond math class, I've come to think of rules and examples as being the two main building blocks for conveying intentions and they correspond to the first and second rows of our 2 × 2 grid from chapter 1. *Rules* are complete and precise, but also abstract, hard to construct, and easily misunderstood. Code can be like that, too. *Examples* are concrete, specific, vivid, and relatable, but they require a leap of logic to fill in missing details. Essentially, if I try to get an idea across to you through examples, I'm asking *you* to find the rule because I'm too lazy to make it myself.

Yet that's how data is used in machine learning. Rules and examples are complementary ways you can describe things, and code and data are complementary ways you can describe *how to do* things.

In the previous five chapters, we've looked at a variety of structures for expressing tasks and how to convey them via code and data. Chapter 3 covered command sequences and how machines can infer them from data using unsupervised learning. Chapter 4 covered conditionals and how machines can infer them from data in the form of decision trees. Chapter 5 covered transforming values through the use of variables and how machines can learn to do the same thing from data in the context of neural networks. Chapter 6 covered loops and how machines infer looping structures through reinforcement learning applied to the data that comes from interacting with an environment. Chapter 7 covered defining new functions and how machines can create and exploit them through convolutional networks applied to visual data. The lesson here is that we have a choice: we can convey each of these structures to machines using code or data, whatever works better for us or for the task at hand.

This chapter focuses on the fact that the code versus data choice is a false one and we should really be using code *and* data. Instead of saying "{1, 4, 6, . . .}," I can say "natural numbers that aren't prime, such as {1, 4, 6, . . .}." Combining the two provides the concreteness of examples with the generality of rules. It gives you two chances to get the details right. You might have trouble remembering whether the natural numbers start with zero or one, but the example makes that clear. You might have trouble knowing how to continue the pattern based on examples, but the rule makes that clear. As a listener, the two modes of communication support each other to solidify your understanding. As the speaker, combining code and data increases the chances of getting your intentions across. Code and data are the peanut butter and chocolate that make up the Reese's Cups of telling machines what to do. As delicious as both are alone, there's something really magical that happens when you put them together.

Pretty much every concept, big and small, that appears in this book was served up with both a verbal description and examples. Chapter 4 introduced the major idea of an *if-then-else* pattern, then illustrated it with examples from weddings. Chapter 6 mentioned in passing that infinite loops are used in "any computational device" and went on to give examples to clarify what counts as a computational device: "a cell phone, a video arcade machine, a climate control system, a network router, a car." This very paragraph describes how verbal descriptions and examples are often used together, then provides a few examples of combinations used in this book. It's exampleception!

Given their prevalence in daily life, it seems only natural that we'd tell machines what to do using combinations of code and data. Unfortunately, there really aren't yet systems in common use that work that way. But the computer science research community is exploring various possibilities, and a few promising ideas are emerging: We'll look at how other people's data can be used to help you generate code, how *your* data can be used to help you generate code, and we'll explore how data and code together might someday let you tell machines what to do more effectively than either alone.

But first we'll look at a few more examples of how we combine data and code to help us understand each other better.

DISHING ON WASHERS

My wife and I were living in a rented apartment in Atlanta while I was working on this chapter. Every twenty-six minutes, a distracting beep came from the kitchen. I couldn't tell whether it was coming from the dishwasher, the washing machine, the coffee maker, the microwave, or the part of my brain that looks for excuses to take a break from writing.

I wanted to rule out the latter, perhaps by succumbing to it, so I (super productively) looked up the model number of the dishwasher and found the owner's manual online. The document overflows with great examples of how people convey complex tasks to each other, drawing on all the ideas we've talked about in the book. It also combines bits of codelike rules with datalike examples to help make the message clear.

Here's a passage from the manual on how to load the bottom rack of the dishwasher:

The lower rack is best used for plates, saucers and cookware items. Large items, such as broiler pans and baking racks should be placed along the sides of the rack. If necessary, oversized glasses and mugs can be placed in the lower rack to maximize loading flexibility. Plates, saucers and similar items should be placed between the tines in the direction that allows the item to remain secure in the rack.

The manual is providing high-level guidance, commands telling you how to arrange things (as in chapter 3), and a conditional "*if*" (as in chapter 4). The phrase "to remain secure in the rack" is really interesting because it's not an explicit instruction that you can follow. Instead, it conveys an *objective*—arrange the items so they remain secure. If you run the dishwasher and get some data that indicates that an item has fallen over, you should change what you do next time. I see this style of writing as treating you as if you were a reinforcement learner (chapter 6), giving you some leeway in what you do while providing you a metric for how to assess success. It's the *explain* box from the 2 × 2 grid in chapter 1.

Next, the manual presents a warning:

IMPORTANT: It is important to ensure that items do not protrude through the bottom of the rack or the silverware basket where they will block the rotation of the lower spray arm. This could result in noise during operation and/or poor wash performance.

Again, you get an instruction to keep the item from protruding, but also by implication a way to measure success—you'll know if something has gone wrong because of the noise and dirty dishes that result. As one last assurance to help you get it right, there's a diagram right next to the text showing an example of dishes neatly loaded and the lower spray arm visible beneath the rack, cheerfully ready to rotate freely. The example isn't something that you should replicate precisely in your own dishwasher; you probably don't even own these exact dishes. Instead, it illustrates an example of an idea or rule that you can apply to your own dish arranging.

Other sections repeat this strategy, combining abstract instructions and concrete examples to paint a complete picture. In this next excerpt, we get a category rule, and then training data to make sure you understand what sort of thing the phrase refers to:

The silverware basket may also be used for small items, such as measuring spoons, baby bottle nipples, plastic lids, or corn cob holders.

Without the examples, you might not know what counts as "small" —or as "items," for that matter. Without the descriptive phrase "small

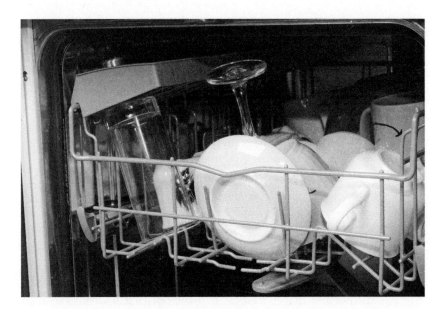

Figure 8.1 A dishwasher should be loaded so that the spray arm is able to turn without hitting any of the dishes. *Source:* Photo 28762187 ©Kaparulin | Dreasmstime.com.

items," you might find it tough to figure out what baby bottle nipples and corn cob holders have in common. (Stuff that my parents were embarrassed for us to ask for when visiting other families?) Combining the examples and the descriptive phrase packs a one-two punch that greatly increases the chances the reader will understand what's going on.

The manual also combines abstract instructions and measurable objectives so you can evaluate your own experience to know whether you are on track:

Avoid allowing items to nest together, which may keep them from being properly washed.

It is saying "do this" and "here's why." Either one alone doesn't quite cut it. If it just said "avoid allowing items to nest together," you might not know how much effort to put into it. Should you design a little demilitarized zone around each bowl? Only wash one spoon at a time? The "keep them from being properly washed" provides the necessary context—if the stuff is coming out okay, you're fine. On the other hand, if the manual just said "make sure all items are properly washed" without the how, you'd be justified to yell at the appliance, "Don't make it my responsibility to properly wash dishes. That's literally your *one job!*"

This technique of providing instructions and stating evaluation objectives to make sure you are following them reasonably can be combined with looping structures, as in chapter 6:

Inspect and clean the filters periodically. This should be done every other month or more depending on usage. If there is a decrease in wash performance or dishes feel gritty, this is also an indication the filters need to be cleaned.

(The phrase "wash performance" brings to mind sitting in the laundromat watching the clothes go 'round and 'round. Although, I suspect this passage is less an instruction than a legal disclaimer. I didn't even know dishwashers *had* filters.)

Remember that the reason I started down this road was to try to stop the periodic beeping. I did find some instructions for "reprogramming" how the dishwasher makes sounds. It turns out that it's configured to beep twice at the end of the cycle, but you can tell it not to behave that way:

To turn off the double beep indicator (or re-activate it if it was previously turned off), press the Heated Dry pad 5 times within 3 seconds. A triple beep will sound to indicate the end-of-cycle beep option has been turned on or off.

The instructions ("press") are paired with a success measure ("triple beep") so that you can modify your behavior to what the dishwasher expects. Indeed, I found that this evaluation metric was essential. Getting the right rhythm for pushing the "Heated Dry" button is not as easy as it sounds here. If you do it too fast, the button presses themselves interfere with the "triple beep" and you can't tell if you did it. If you do it too slowly, you are simply toggling the "Heated Dry" option on and off and the end-of-cycle setting won't change. Although the manual doesn't say so, it turns out that there's more information in the "triple beep" itself. It either plays the first three notes of the *2001: A Space Odyssey* theme really fast or the same three notes in reverse order, depending on whether you just turned the end-of-cycle indicator on or off. Unfortunately, it doesn't tell you which is which. Even more unfortunately, none of this is related to the beeps I've been hearing. Two steps forward, three beeps back.

The "Troubleshooting Tips" section of the manual makes good use of trigger-action programming (chapter 4), providing a list of problem triggers like "Black or gray marks on dishes" and the corresponding actions you should take, such as "Remove marks with a mild, abrasive cleaner." (Is that an oxymoron?) This section also takes advantage of functions (chapter 7): Instead of the manufacturer explaining the same procedures over and over, you are sent to another place in the manual that includes a thorough discussion. For example, the suggested resolution to "detergent left in dispenser cups" is:

Reposition the dishes, so the water from the lower spray arm can flush the detergent cup. See the *Loading* section. Reposition dishware that may be keeping the dispenser cup door from opening. See the *Loading* section.

And the "Loading" section provides step-by-step instructions for loading the racks. These functions are paired with useful information about what they need to accomplish (making sure the spray arm or dispenser isn't blocked), again combining code and data elements together.

The troubleshooting section of the manual even uses a little recursion! For resolving "cloudiness on glassware," it suggests:

See "A white film on glassware, dishware and the interior" in this *Troubleshooting* section.

A section referring to itself, spotted in the wild! David Attenborough would be so proud!

Variables (chapter 5) help make sure the same instructions can be applied in different scenarios:

The Delay Hours feature Delays the start of a wash cycle up to 8 to 12 hours depending on model. Press Delay Hours, to select the number of hours, then press Start one time and close door to begin countdown. The cycle will start once allotted time has expired.

Here, "allotted time" stands in for whatever delay you chose—a variable.

All in all, the manual, twenty-four pages long and repeated in three languages, provides great examples of how to combine code and data to convey tasks. Pretty productive after all. However, I couldn't find anything about why I'm hearing a beep every twenty-six minutes. Hmm, maybe I should spend a few hours reading about noise-canceling headphones.

TALKING MACHINES

Product manuals and employee handbooks and medical discharge instructions and the like are documents written by people to convey multistep procedures to people. In my experience, they consistently combine examples and rules. Given how universal and maybe even essential this combination is for easily and accurately conveying tasks, it's surprising that computer systems don't use it: it's data or code, but not both. Some initial examples are out there, though, that might help us see what future programming systems might look like. The rest of the chapter is devoted to highlighting a few promising directions.

The Scratch programming language from MIT is designed to give beginning coders a platform for making creative and powerful programs. I highly recommend it, as it encourages you to play around and build interesting systems in a fun, colorful, nonthreatening way. The current version includes extensions that let you write programs that use face tracking (got your nose!), automatic machine translation (*c'est magnifique!*), text to speech (I'll say!) and more, all built on top of powerful machine learning models trained with massive amounts of data. The Teachable Machine project goes a step further and provides a relatively simple interface for providing your *own* data to create customized machine learning models for doing things like identifying items in pictures. The recognizers can then be incorporated into other programs you write in Scratch, and even in other languages. I find these projects really exciting because they let you, independent of your past programming experience, get the benefits of advances in machine learning more or less by just supplying data machine learning can learn from. The data is ingested by functions that do all the learning work, and the result is packaged as a function that can be used just like any other. Here, the data *becomes* code, which is awesome. But you aren't using code and data *in tandem*, so you're just getting a fraction of the power that comes from conveying tasks using their combination. There have been some examples of integrating code and data more closely. But to talk about the first one, we need to talk a bit about programs that talk.

Chapter 3 described unsupervised learning—examining data to find useful statistical relationships without being told ahead of time how the

data clusters. Chapters 5 and 7 discussed deep learning. One very exciting development in recent years has come from combining these ideas—
training deep networks using unsupervised data specifically to process
language.

There are a lot of tasks people would like to automate that require the
computer to have some degree of "understanding" of words. For example,
imagine you are planning a graduation party. You could use help comparison shopping for venues and responding to requests from relatives
about last-minute logistics. In both these cases, essential information is
expressed as sentences: descriptions of venues on a web page that you'd
have to read to extract key facts, or questions coming in via text about
parking or gift policy or whether the event will be livestreamed. It would
be nice to be able to create a trigger-action program as in chapter 4 that
says "if you receive a text about what people should wear, then send them
back the sixth answer in the FAQ." But determining if a message is about
dress code means the program has at least some minimal skill at reading comprehension, including knowing that a question about a "monkey
suit" is about clothes and not about whether people can bring pets.

There are no programs that can be said to *understand* what they read.
Not yet. But programs that can accurately identify the category a passage
of text belongs to have been around and steadily improving for thirty
years or so. Today, such programs are helping to turn on and off the right
lights by voice command, answer questions about the weather through
Siri, and correctly route customer service requests over the phone. Match-
ing a sequence of words against a fixed list of possible interpretations is
a problem that's analogous to recognizing pictures of animals in images,
as we discussed in chapter 7. So, not surprisingly, neural networks similar
to the ones that drastically improved image recognition have also revolu-
tionized text interpretation problems.

The journey from early card catalog systems that let you type keywords
to find books to today's broader and more insightful language processing
goes through Antarctica. Oh, sorry, I meant autocorrect. Not sure what
went wrong there. As I'm sure you (will/are/can) have seen, autocorrect,
and the related idea of predictive (text/of/to) text, uses information about
what you just typed to guess what word might come (from/to/up) next.
These systems have been getting steadily better at their job as machine

learning approaches allow longer and longer contexts to be taken into account when making predictions. Modern approaches make effective use of the last 256 words to guess what's coming. That's a decent amount of context, and it means that these systems are much more likely to accurately guess what's next. To train a predictive text system, we do not need to put any effort into collecting specific judgments from experts. Instead, all we have to do is gather mountains of naturally occurring text from online sources. Every single word in the collection becomes its own mini-puzzle: guess that word from what comes before it and find out immediately whether you were right! Because no explicit human annotations are needed, this kind of training is considered a form of unsupervised learning.

At the heart of any predictive text system is a *language model*, a statistical estimator or a neural network that can assign probabilities to words in context. Researchers from several of the big AI powerhouse groups have worked out ways to train enormous neural networks with tens of *billions* of weights using hundreds of billions of words of text. These models can cost many millions of dollars to develop and train, so each one has its own following and even its own name. For example, there's ELMo (Embeddings from Language Models) from the Allen Institute for Artificial Intelligence, BERT (Bidirectional Encoder Representations from Transformers) from Google, and GROVER (Generating aRticles by Only Viewing mEtadata Records) from the University of Washington. I find it poetic that these projects are elaborations on top of a more primitive language model called GloVe (Global Vectors) from Stanford. After all, what is a glove but a very basic Muppet?

Some of these networks, such as GPT-3 (Generative Pre-trained Transformer 3), OpenAI's boldly non-Sesame-Street-named model, have been available for people to play around with online. One remarkable property is that they are good at far more than just predicting the next word. Just as image recognition networks create their own internal feature representations that are useful for visual problems beyond what they directly trained for, language models built for predicting the next word can be applied to all sorts of language tasks, including the FAQ problem mentioned earlier. Google Search uses BERT's representations to better match the queries you type to relevant web pages, for example.

Similarly to how convolutional networks can be used in new problems such as monster spotting by reusing the internal feature representations they built for other tasks (chapter 5), the networks in language models can also be incorporated into other models to solve new tasks. But, because language models are adept at finding patterns in text, they can be reused in a way that doesn't involve retraining them. Instead, you can directly leverage their ability to fill in the blanks to tell you what they know. That is, you can essentially just *ask* them.

To illustrate this idea, I made a list of ten colors (white, black, yellow, red, etc.), and for each one, I thought of ten things that are that color (steamed rice, the clothes you wear to a funeral, a rubber duck, Rudolph's nose, etc.), resulting in a list of one hundred item-color pairs. I signed up for access to GPT-3 and asked it to complete a series of statements such as "The color of a Smurf is. . . ." In about two-thirds of cases, including all the examples above, GPT-3 came through with the color I was expecting. For a system that can't see and is only exposed to language that it doesn't actually understand, it knows an awful lot about the colors of things in the world. When it didn't come back with a color, it came back with reasonable if unhelpful completions, such as:

The color of a pencil is . . . not important.

Or

The color of an egg yolk is . . . determined by the hen's diet.

It occurred to me that I hadn't really conveyed what I was looking for. Maybe GPT-3 would be sensitive to being told what to do through examples. I prompted it with a few correct answers first, then the query, like so:

The color of a lilac is purple
The color of Cookie Monster is blue
The color of blood is red
The color of snow is white
The color of a tomato is . . .

These additional examples helped a lot! GPT-3 now gave one-word color names as its response for 100 percent of the items, and 92 percent of them matched my choice. For comparison, I gave the same one hundred questions to my two adult children and they matched my choice 98 percent of the time. I think that's pretty remarkable. For all three of them, really.

As people have played around with GPT-3 to get a sense of its view of the world, other amazing feats have surfaced. By feeding its prediction back in as an extension of the text that you are giving it, you can get GPT-3 and other language models to complete the text you give it. It's like a game I used to play on long family car trips. Each person in the car takes turns adding a word to the end of a sequence. Basically, the participants write a story, and no single person in the group knows where it's going to go. Using this idea, if you prompt GPT-3 with something that looks like a movie script, it will write the rest of the scene for you. If you prompt it with something that looks like a newspaper article, it'll complete the report. And if you give it something that looks like code, it will write a program on your behalf. Hmmm, that could definitely come in handy to help people tell machines what they want them to do!

CODING MACHINES

Encouraged by the coding prowess of its language model GPT-3, OpenAI teamed up with GitHub, the largest online code repository. GitHub is a subsidiary of Microsoft, which provides backing for OpenAI as well. Together, they created Copilot, a version of GPT-3 trained on code uploaded to GitHub and specifically modified to produce valid programs.

The makers of Copilot bill it as an "AI pair programmer," but that's not quite right. In pair programming, two coders work together at a workstation. One writes the code, focusing on tactical, local aspects of how to express necessary ideas to the computer. The other checks what's being written and plans strategically for ways that the code might be organized more effectively. Or gets coffee. Whatever helps.

In contrast, Copilot looks more like a jazz musician to me. You start writing your code, and Copilot riffs on it, guessing where things could go and suggesting exciting new directions. Sometimes it seems like Copilot is reading your mind—writing a flawless function just based on the name you started typing. Other times, it acts like a crazy uncle who got set off by something you said and is now off ranting about it in his own private universe. If a real-world copilot acted like Copilot, air travel would be a lot more . . . adventurous, for one thing.

With the help of one of my undergraduate students who had been using it to be more productive when programming, I played around with Copilot a bit. It's adept at all the things we've been talking about. It writes code with commands, conditionals, and loops. It uses variables and writes functions, choosing remarkably reasonable names for them as it goes. I had it write a word-counting program, as in chapter 7, which it did nicely. Then I decided to give it a shot at the rainfall problem.

The rainfall problem, introduced in chapter 2, was a simple programming task given to beginning computer science students in 1981. They were asked to code a function that would read in numbers, one at a time, stopping when it read the number 99999. At that point, it was to output the average of the numbers it had read. About 60 percent of beginning students and even intermediate students flubbed this task, which many people have taken as evidence that programming is tricky and we need better ways of teaching it. I mean, sure.

What if those students had had access to Copilot? Would programming no longer be so tough? I started to write a function in Copilot called "rainfall_problem_original" and included, as a helpful comment in the code, the original statement of the rainfall problem given to the students in the experiment:

Write a program which repeatedly reads in integers until it reads the integer 99999. After reading 99999, it should print the correct average. That is, it should not count the final 99999.

Here each of the integers (positive or negative whole numbers) is supposed to represent the amount of rain that fell on a given day. The fake rainfall number 99999 is a marker that the list is over and it's time to compute the average.

Copilot helpfully wrote Python code that it predicted would follow such a comment. Since you probably don't know Python, looking at the literal code is maybe not that helpful. But I can translate it to the style we've been using in this book to get across the main ideas. It wrote:

```
set total to 0
set count to 0
```

```
do forever:
  read a value in and assign it to the variable rainfall
    if rainfall is 99999:
    break out of the forever loop
  else:
    add rainfall to total
    add 1 to count
print out total / (count-1)
```

What do you think? Is programming easy now? Is my work here done and I can let Copilot figure out that beeping noise coming from my dishwasher? Well, let's see. First let's look at some of the things Copilot did well. It did a nice job naming the variables. It used "*rainfall*" to represent the values being typed in, which indeed are supposed to be the amount of rainfall that was measured. It chose the name simply because I named the function "rainfall_problem_original." Crazy, right? It has seen enough about the way programmers name variables and functions that it could conclude that using *rainfall* as the input value was a great guess. It chose the name "*total*" for a variable that keeps track of the total rainfall so far and "*count*" for a variable that keeps track of how many rainfall entries have been received.

The looping structure it chose is pretty interesting. It used a forever loop, but in James Bond–like *Never Say Never Again* style, it jumps out of the forever loop once the 99999 signal is received. It's like, "I'll be with you until the end of time. Wait, was that a 99999? Never mind, I'm out of here." Getting this looping structure right was one of the main difficulties the student programmers had. It's not a typical *while* loop because we need to do part of the inside of the loop (reading in the number) but not the other part (adding to the count and total) when a 99999 is entered. There are a handful of ways of expressing this idea, and Copilot chose one that works perfectly well.

All good so far. The program totals up the rainfall and counts the number of values that went into the total. Those are the two things we need so we can report the average. It stops when it gets a 99999, breaking out of the loop without messing up the total or count. Finally, it's time to print out the average rainfall. Copilot decided to divide the total by count minus one. That is not right. It should be just the total divided by the count. My guess is that the admonition to "not count the final 99999" caused some confusion. So close, though!

To me, the lesson here is that Copilot is not an automatic programmer where you can just explain what you want and let it run. Copilot helps you write code, but you are still responsible for making sure that code is correct. It uses machine learning to create and implement a strategy, gleaned from training on tons of human-written code, to help you create your specification. Then you need to check that what it suggests is what you actually want. If Copilot is the copilot, you are still the captain!

DO YOU WANT TO BUILD A PRO-GRAM?

Deep inside the computer, drawings are just code. For example, a simple snowman might look like this:

```
draw a circle 3 units over 1 unit down
draw a circle 3 units over 3 units down
draw a circle 3 units over 5 units down
```

We could express that same idea with a variable, as in chapter 5, and a loop, as in chapter 6:

```
for i from 1 to 3:
    draw a circle 3 units over 2*i-1 units down
```

The clever bit here is the formula two times i minus one, which you may remember from the odd numbers example at the start of the chapter. It takes on the value one when i is 1, three when i is 2, and five when i is 3. The result is the same three circles that were drawn in the first example, but in a slightly more rulelike form.

There are some advantages and disadvantages to making drawings using code. An advantage is that the structure of the picture itself is made explicit. So, if we want to make our snowman a little taller, a very tiny change will do the trick (changing the 3 to a 4 in the definition of the loop):

```
for i from 1 to 4:
    draw a circle 3 units over 2*i-1 units down
```

Now there are four snowballs stacked up. We could even make the snowman a *lot* taller by changing the 4 to a 10:

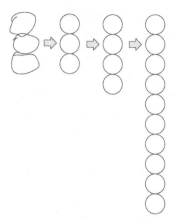

Figure 8.2 A snowman can be hand-drawn or programmatically generated.

```
for i from 1 to 10:
    draw a circle 3 units over 2*i-1 units down
```

The disadvantage is that the code doesn't *look* like a snowman. Besides, who ever heard of a snowman ten snowballs high?

Graphical interfaces for making drawings are sometimes referred to as *whizzy wig*, spelled WYSIWYG, for "what you see is what you get." A great thing about WYSIWYG is the tight link between what you click and drag and what it visually produces—no imagination necessary. When coding, on the other hand, there's a mismatch. You only get something you can see after running code: YOGSYCSARC. And *yog-sicks-ark* doesn't have as nice a ring to it.

To get the directness advantages of WYSIWYG and the generalizability advantages of YOGSYCSARC, it would be great if drawings could be turned into code and then modified. Ideally, you could make changes to either form—tweak the code, and the picture changes; alter the picture and the code updates to match. There's something rather Lamarckian about this idea: Stretch the giraffe's neck and the DNA is rewritten to match.

Perhaps not surprisingly, no one has put together an interface like this. But a research project by some of my colleagues at MIT and Brown helped establish some of the basic ideas that might make it possible. In their project, people can hand-draw sketches of the kind you might find in a scientific paper—circles and boxes connected by lines and arrows. The

computer will automatically turn these sketches into sequences of drawing commands as in the three-circle snowman example at the beginning
of the section. It then performs a deeper analysis and produces high-level
code as in the *for* loop example. It can identify symmetry and repetition
in hand-drawn diagrams, and even correctly use conditionals (as in chapter 4) to handle exceptions and boundary cases. It's very slick.

Discovering the structure in a sketch is an interesting problem. Let's
say you hand draw four squares in a square, like a window with four
panes. When you draw your squares, they will not all be identical. If the
system *makes* them identical, it is technically introducing error. But if it
doesn't make them identical, it may be missing the *intent* behind your
drawing. So it makes sense to strike a balance between the simplicity of
the resulting code and the amount of mismatch between the output of
the program and the hand-drawn version. This simplicity-versus-accuracy
trade-off is central to all applications of machine learning. In terms of the
machine learning recipe introduced in chapter 3, this balance is generally
handled by setting up the loss function to put weight on both avoiding
complexity *and* avoiding error.

In a sense, the system is learning a drawing program from a single
drawn example. However, the learning process itself is aided by more
traditional machine learning in multiple ways. Hand-drawn circles and
boxes and lines are turned into the appropriate drawing commands using
a machine learning convolutional network, of the kind discussed in chapter 7. This network is trained on other people's images to help it effectively
translate your sketch into code. The system also exploits machine learning to speed up the drawing-command-to-drawing-program generation
process, using data from past successful searches for programs to better
invest its efforts in the future. The result accelerates the program-creation
process from about an hour to about a minute. However, even a minute is
not really what you want if you are using a drawing program. People find
delays longer than ten seconds distracting enough that they will start
working on something else instead of waiting: "Maybe I'll check Facebook right quick." (Arguably, another shortcoming is that it would suck
the life out of Pictionary; the all-too-perfect circles and boxes don't leave
your teammates anything to complain about.) So, for now, this technology remains lab-bound.

There are other projects that share this idea of using your data to help you create code. In chapter 4, we talked about trigger-action programming and how it can be used to automate "smart home" behaviors such as turning lights on and off. Trigger-action programs are one of the easiest kinds of code to write, but data can still be quite useful in making it better.

Here's a concrete scenario. In your office, you have a desk lamp and various sensors for detecting information such as whether there's been movement recently, whether the office door is open or closed, and the temperature and the light level in the room. You don't love to waste energy, so you'd like the lamp to shut off when it's not being used. A trigger-action program would do the trick:

- **if** the lamp isn't being used: <u>turn off the lamp</u>

But, of course, there's no such thing as a sensor for whether or not the lamp is being used. The best you can do is to use a proxy based on the available sensors. If the door is closed and there hasn't been motion in a while, that's a pretty good signal that you're gone, right? But how long is "a while" in this context? Is ten minutes too long? Is two minutes too short? Data can help answer this question.

Some colleagues of mine at the University of Chicago (including Blase Ur, who we met in chapter 4) built an experimental system with this idea in mind. They instrumented the offices of some volunteers with sensors and lamps, collecting data over the course of a week or so that included when the participants turned their lights on or off. The data relating sensor information to switching the light on or off was fed into a custom machine learning system, which found trigger-action programs that would mimic, as closely as they could, the light-switching decisions of the office occupants. A list of the most accurate rules found was presented to the participants, who were asked to choose the rule they would like to have running in their office.

The results were really interesting. For one thing, the rules the system wrote varied from participant to participant. That's because the sensors were arranged differently in each office and, of course, people have different habits for how they use their lights. That suggests that having some central authority decide on the best rule for the entire building isn't a great idea. It would be like having everyone in the building as your roommate, with the lights following their preferences instead of yours.

In addition, the precise sensors to check and durations to wait would have been difficult to guess in advance. The system proposed rules like

- **if** <u>the door has been closed for</u> 5 minutes
     and <u>there has been no motion for</u> 2 minutes:
  <u>turn off the lamp</u>

and

- **if** <u>the time is</u> 7:20pm: <u>turn off</u> the lamp

In retrospect, it's possible to figure out why those particular times are good—some people leave the office on a schedule to catch public transportation, for example. The data is adding some real value here because it's easier to recognize a correctly specified pattern than it is to come up with one from scratch. Because the data capture more about regularities in your behavior than you may be consciously aware of, it's tempting to suggest that we'd be better off just letting the machine learner take over every one of our decisions. After all, as many tech giants learning from our data have said, it knows what you do better than you do.

That's true, but it's still important for people to make the decision about what behavior they want delegated to the machine. As a commonsense example, office dwellers turn their lights off right before they leave. Is that the behavior they want to automate? Perhaps not. It makes more sense for the lights to go off *after* they leave, especially if the alternative is for them to grope around wildly from the desk to the door in the now darkened room. Instead of having the lights reflexively copy that behavior, people should get a chance to double-check that the rules capture the behavior they want.

That's what the office automation study found. Given a set of rules of varying accuracy, participants often chose ones that the system didn't think were the most accurate. People have preferences, and translating possible behaviors into code that they can vet and approve if appropriate is a great way of respecting those preferences and letting them maintain their autonomy.

LOST WITH A COMPAS

Using your data to generate home automation rules that you can review, edit, and then deploy starts to show some of the benefits of combining code and data. In particular, the code is better than what you'd write on your own, and the behavior is closer to what you really want than a purely statistical machine learner would choose for you. There's a terrific synergy that comes from combining them, as good appliance manual writers know. But turning data into readable code has other advantages that I think are worth reflecting on.

Rules are a key fuel on which society runs. They are intended to keep us safe and let us work together better. But they also limit our freedom and have an impact on our lives in ways small and very large. A rule like "You must be at least 54 inches to ride The Viper roller coaster at Six Flags" can make or break your day. It was, I hope, written by well-meaning people trying to create a safe environment for everyone. I mean, I don't think it's based on data collected from letting different-size kids onto the roller coaster and seeing which ones were injured. Eek. But, increasingly, machine learning is being used to create rules that affect people in more consequential ways. Banks, for example, have an incentive to analyze past data about which customers have repaid loans on time and in full to come up with rules for deciding who should be considered a safe credit risk. A big difference between the Viper ride and a machine learning–based loan rule is that you know why you were refused entry to the Viper. With a bank loan, it might just be "because the computer said so." And hiding behind that decision could be some unethical practices.

In 2016, just as broad interest in machine learning was beginning to heat up, the investigative publishing group ProPublica released a report with the title *Machine Bias*. It provided their analysis of a widely used but little-known piece of software called COMPAS, Correctional Offender Management Profiling for Alternative Sanctions. Many, many judges in the US choose, or are sometimes required, to use the software to get a dispassionate evaluation of whether a criminal defendant is likely to become a repeat offender. The program takes in a set of over a hundred features that describe the defendant and spits out a risk score saying whether their probability of being arrested again is low, medium, or high. The results are used to decide on bond amounts or even how long a sentence the defendant will have to serve.

The ProPublica team gathered a data set consisting of risk scores for over seven thousand people arrested in 2013 and 2014 in Broward County, Florida. Since the report was written in 2016, ProPublica followed up all these people to see whether they were charged with new crimes in the two years that followed the earlier arrest, something COMPAS is designed to predict. They found that the system was moderately accurate, correctly predicting the defendant's future behavior around 60 percent of the time. They further found that this accuracy was similar for both white and Black defendants. But, worryingly, the errors it made were in very different *directions* for the two groups. When the system predicted incorrectly, it was more common for it to predict that a non-reoffending Black person was high risk and to predict that a reoffending White person was low risk. That is, the system appears to be much more lenient on White defendants and harder on Black defendants.

Also, importantly, there is ample evidence that Black defendants are more likely to get caught or blamed for reoffending than White defendants as a consequence of a higher level of policing in Black neighborhoods and a biased criminal justice system. The algorithm is designed to predict arrests, which are measurable but unfair, and not crimes, which would be much a more desirable prediction goal. Once again, we see that defining good objective functions is hard and that our choice of proxy can have significant repercussions.

This report spurred a great deal of discussion and controversy and self-reflection, especially among machine learning researchers. Although the set of features COMPAS uses to make its predictions do not include explicit racial markers, and its proprietary decision-making was probably not designed using machine learning, the findings vividly raise the prospect that machine learning systems can discriminate. It helped spur the creation of a vibrant new field known as Fairness, Accountability, and Transparency (FAccT), which studies how to make machines that we can trust to make decisions about people.

Personally, I think it's great that ProPublica shined a light on the ways machine learning can treat people unfairly. It helps reveal that all the parts of the machine learning recipe—the way rules are represented, the way the loss function scores them using data, and the way a good rule is ultimately selected—can contribute to unequal treatment. Just because machine learning is math doesn't mean it can't reinforce stereotypes and adopt societal biases.

Despite the positive outcome for the field, the COMPAS result itself turned out to be misleading. Apart from ProPublica's overall concern that scores representing the risk of rearrest themselves carry significant risks by giving algorithms undue "control" over human lives, the organization lodged two concrete complaints against COMPAS. Specifically, COMPAS is more likely to erroneously predict (1) that Black defendants would be future reoffenders and (2) that White defendants were low risk. It's like the system undertrusts Black people and overtrusts White people, which is just the sort of thing you'd expect from a racist computer system. Not good. And just like when you hear a kid casually using racial epithets, it's the *parents* who should be blamed—the makers of COMPAS, in this case.

So let's daydream that the designers of COMPAS apologize for their error. They go back to the drawing board and create a new version of their software. Of course, it still makes mistakes—perfect crime predictions are only possible in the movie *Minority Report*. But now COMPAS makes mistakes in similar directions for both groups of defendants: If the system says a defendant is at low risk for reoffending, the likelihood that the system is wrong doesn't change when you consider the person's race. Same for people the algorithm claims are high risk. It's consistently 60 percent accurate. Outstanding.

The staff at ProPublica is ready to celebrate the more just algorithm they inspired when they realize that something is off. In the original system, when the algorithm said a defendant was "low risk," the likelihood of a rearrest was around 30 percent, regardless of race. When the algorithm said a defendant was "high risk," the likelihood of a rearrest went up to around 60 percent, regardless of race. But the newly "fixed" system applies a different meaning of "low" and "high" for Black people and White people, with more Black people being rearrested in both categories by about ten percentage points. That means a Black defendant labeled "low risk" is more likely to be rearrested and a White defendant labeled "high risk" is less likely to be rearrested. COMPAS moved the goalposts in a way that looks pretty unfair to White defendants. I mean, come on,

Figure 8.3 A snowman can use a compass. *Source:* Image created by author with DALL•E 2 (OpenAI).

COMPAS. Must you always be biased against one of the groups? Even in our daydream?

Unfortunately, the answer is yes. There are multiple ways of defining what it means to be fair. Mathematically, many of these notions conflict with each other, meaning that an assessment that is fair by one measure is *necessarily unfair* by another. And the notion of fairness suggested by ProPublica's analysis and the one in the daydream are examples of conflicting assessments of fairness. Getting things right is complicated. In 2014, Attorney General Eric Holder expressed a concern that automatically assigned risk scores may endanger our ability to "ensure individualized and equal justice." It's an interesting choice of phrase. Arguably, "equal" and "individualized" are at odds with each other, as one word encourages comparisons across people and the other discourages it. These tensions are real and require us to repeatedly reengage. We can dream that societal discord can be automated away, but there's no substitute for discussion and debate.

Instead of setting ourselves the impossible task of finding the one true fairness concept and making our algorithms adhere to it, maybe we should just be as transparent as possible about what the algorithms are doing. If we can look at the rule, we can see if it says "I don't trust Black people" or if, instead, it is making a reasonable assessment and the unfairness isn't in the assessment rule but in other sectors of society that result in large numbers of Black people getting arrested. Let's be clear on how the rules work and then debate whether we should follow them.

The idea that we should make the output of machine learning systems easier to understand is itself controversial. Geoff Hinton, a neural networks pioneer and winner of computer science's Nobel-like Turing Award, tweeted:

Suppose you have cancer and you have to choose between a black box AI surgeon that cannot explain how it works but has a 90% cure rate and a human surgeon with an 80% cure rate. Do you want the AI surgeon to be illegal?

Cynthia Rudin, winner of the much less impressive-sounding but actually quite impressive Squirrel AI Award, has a killer comeback: It's a false choice. For every machine learning problem she has encountered, she has been able to find an interpretable rule that performs just as well. When we're deciding between an AI surgeon that has a 90 percent cure rate that can't explain itself and one that has a 90 percent cure rate and *can* explain itself, the choice is clear. Putting her money where her mouth is, she reanalyzed the ProPublica data set and found a simple rule that acts a lot like the proprietary COMPAS algorithm. Written as a conditional expression of the sort we saw in chapter 4, the rule is:

```
if age is 23-25 AND priors is 2-3: predict high risk
else if age is 18-20: predict high risk
else if age is 21-22 and sex is male: predict high risk
else if priors > 3: predict high risk
else: predict low risk
```

This rule looks quite sensible to me. If the defendant is young, a repeat offender, or male, there's reason to be worried. There's nothing explicitly race-related in there, although it's justifiable to be concerned that the reliance on priors might affect Black people more strongly because the Black community is policed more heavily. My point is not that this rule is the best rule. It's that an explicit piece of code like this, learned from data, is a much more reasonable starting point for discussing public policy than a rule that may be just as fair but you have no way to tell because you can't read it. Once again, rules and data can achieve more together than apart.

A SNORKEL DEEP DIVE

When we tell other people what we'd like them to do, we use combinations of rules, examples, and objectives. The systems we've talked about so far in this chapter have primarily used data to help you make code. I think the path forward is for us to be able to provide combinations of code and data (and goals) and for computers to work out what we need them to do from these complementary hints.

Some of my colleagues at Brown University advocate teaching people to write code that integrates examples, using a method they call *test-first programming*. Students are encouraged to write code like

```
define the function count positives in list:
    making sure count positives in (1, 5, 2) returns 3
    making sure count positives in (99999, -1, 6) returns 2
    making sure count positives in (11, 0, 1, -14, 4) returns 3
    making sure count positives in () returns 0
    set count to 0
    for each number in list:
        if number is at least 0:
            add 1 to count
    return count
```

The idea here is that every function they write comes along with a teeny bit of data. It's not enough data for a machine learning approach to learn what the function should do, but it's useful for vetting the code that's written. If we run this code, for example, it'll tell us that the program produced the right answer for examples one, two, and four, but returned 4 instead of 3 for the third example. The problem is that the code counted zero as a positive number but the example did not. A more powerful system might have used the examples to help fix the code, suggesting the "**at least** 0" be rewritten as "**more than** 0." That would be strictly greater than the systems we have today.

The closest system I've seen to the goal of using both data and code to let people specify tasks is the Snorkel project out of Stanford, which my colleague Steve Bach at Brown worked on. As an example application, imagine you've got thousands of sentences from newspaper articles that mention a pair of people and you want to make a list of which people have had a beef at some point. Here's a few examples of sentences from a hypothetical tabloid-style magazine that mention two people and the word "beef":

1. <u>Kevin Hart</u> and <u>Mike Epps</u> have taken their **beef** to Instagram's comment section.
2. <u>Cardi B</u> has serious **beef** with <u>Peppa Pig</u>.
3. <u>Dwayne "The Rock" Johnson</u> and <u>Vin Diesel</u> reportedly met to squash their **beef** after Johnson called out his "candy ass" costars on Instagram.
4. The celeb chef <u>Rachael Ray</u> created a scrumptious bacon and **beef** birthday cake for dog <u>Isaboo Ray</u>'s 9th birthday.
5. At the Great Texas BBQ Cook-Off, <u>Aaron Franklin</u> and <u>Elizabeth Karmel</u> showed off what they could do with a side of **beef**.
6. At the time, the rapper said that <u>Selena Gomez</u> had **beef** with him whenever he would hang around <u>Justin Bieber</u>, 27, whom she has had an on-and-off relationship with in years past.

The first three are evidence for beefs, but the next three are not. Well, the sixth example actually is about a beef, but not between the two people mentioned in the sentence, so we can't count it.

If you wanted to make a rule that distinguished feud-beef from food-beef, it would be really tough to create one that you'd expect to be highly accurate. You could probably write some decent rules, like

```
if the sentence contains "squash" OR "squashed": guess feud-beef
if the sentence contains "bacon" OR "jerky": guess NOT feud-beef
if the sentence contains "beef with": guess feud-beef
if the sentence contains "with beef": guess NOT feud-beef
  . . .
```

and indeed, these rules are pretty accurate. But all have exceptions, and when the rules disagree, it's not clear which should take precedence. Plus they don't cover all possible sentences.

The machine learning approach to the problem would be to pull out lots of examples and mark each one by hand. Categorizing enough for a machine learning algorithm to construct an accurate rule might be more than you can do on your own. So both the code and data approaches have pretty significant limitations. But Snorkel says, "Why choose?"

Snorkel's approach is to accept any evidence you can provide, in the form of data or code or whatever. You also give it lots and lots of "unlabeled" examples where there's a pair-with-beef sentence but no one has definitively said whether or not it's an example of a feud-beef. Snorkel doesn't take any of these sources of information as definitive ground truth (or ground beef, for that matter) but assesses how all the rules and data fit together, which agree, which are primarily duplicates of other rules, and which seem to provide new insights about the unlabeled examples. It figures out what constitutes strong evidence and what constitutes weak evidence and reasons about what labels are appropriate for the unlabeled examples. Finally, it takes these approximately labeled examples and runs them through a large-scale machine learner, typically some kind of neural network, to identify the most accurate overall rule it can construct.

Impressively, the Snorkel team has been able to show that the system creates state-of-the-art classification systems that outperform what is possible by applying code alone or data alone. That's so cool! I was just saying we need both code and data! Snorkel is considered a "weakly supervised" learning algorithm because it doesn't require gold standard–level labels but can work with a diverse set of hints.

The Snorkel team has honed the design of their system through interaction with end users who are experts in a particular area, but not in programming. For example, Snorkel has helped researchers at the FDA build tools they can use to predict unknown drug interactions, which is awesome and arguably saves lives. The team even held "office hours" every Friday during which they provided informal guidance to beginning Snorkel users from bioinformatics, defense, manufacturing, and other areas. That's right, meetings on their weakly supervised algorithm were supervised weekly.

At this point, Snorkel applications have been mainly academic and industrial, and haven't been honed to apply to more widespread personal uses. But Snorkel's design suggests that future systems could be built that let regular people with problems they want to automate convey to computers what they want, combining code that gets the meat of the problem right with data that can fix the edge cases. And that would make it considerably easier for everyone to leverage the power of computation for themselves.

9

PROGRAMMABLE WORLD
AS YOU WISH

Around fifteen years ago, I created and taught a new introductory computing course and found that convincing students to develop programming skills is a hard sell. Even those who enjoyed learning about computing in a classroom setting didn't see any utility in continuing to program once the semester ended.

I came to see it as a chicken-and-egg problem. If the only things available to program are essentially scientific calculators, and if you don't need to do scientific calculations, why should you learn to program? On the flip side, if people aren't interested in programming, why would anyone make other kinds of programmable things for them? It left me wondering, what if we somehow got through this impasse and programming and programmability were universal? What if programming knowledge were as widespread as computers? How would the world be different?

I get why we don't want some devices open to being programmed by their owners. I don't want to be able to reprogram my car to try to avoid hitting squirrels only to discover that I made a mistake and now the car doesn't slow down sometimes when I step on the brake pedal. It's bad enough that the professionals at car companies goof up that way occasionally. Yikes. But wouldn't it be useful to be able to ask my car to gently alert me if I drive near someplace that sells the dog food I've been looking for? Or to play a loud, angry barking sound if I say the trigger phrase, "Sure, you can have my wallet"?

As the world fills with programmable devices, will we be able to tell them what to do? Or will programming continue to be something reserved for the experts, like landing planes, removing spleens, or, thanks to Steve Jobs, replacing the battery in my iPhone? I believe that a world where almost everyone can program almost all their possessions would give rise to a fundamentally different dynamic, much as widespread literacy has brought us road signs, application forms, birthday cards, comic strips, and so much of the texture of our society.

LIFE HACKS

It's not clear to me how to break the logjam caused by people not being interested in programming and there being not much interesting for them to program. Perhaps imagining a different world where we can all tell machines what to do would help.

It's hard to predict, of course, just as it's hard to predict how people will bring reading and writing to bear on their lives before a community is broadly literate. We can get some hints if we take a historical perspective as we did in chapter 1. We can also look to see what sorts of everyday programming interventions people who can code do today. I've collected a bunch of examples from my own experience and from some of my colleagues'. I also solicited ideas on Twitter, which resulted in a great list. I never thought I'd say it unironically, but: Thanks, Twitter!

One common category is an "early adopter" situation. There are some things that are sufficiently useful that someone will share or sell a high-quality version of them soon. You can just wait for it. But if the need is pressing. . . .

In 1994, my wife and I had our first child. He was born prematurely at one pound and half an ounce. We had recently moved to Rhode Island, and our family and most of our friends lived pretty far away. But the internet was starting to take off, and I had the idea of posting regular updates as web pages they could visit to follow along. After all, Max (a small but strong name for a small but strong baby) was the first grandchild on both sides of the family, and my wife and I were among the earliest in our friend group to have a kid. There was a real demand for information about his situation.

These days, you can throw together a WordPress site or even just post regular status updates on Facebook or Instagram. But 1994 was five years before the word "blog" was even invented. Mark Zuckerberg was ten years old and just learning to program in Atari BASIC. I was more or less on my own. I wrote some simple scripts to convert my diary entries into web pages and even created a pipeline so I could digitize photos and movies. We kept track of Max's daily weights and I wrote a tool that converted the numbers into a graph and linked it to the rest of the site. It was one of those things that served a purpose but also just made me feel a little bit less helpless. At that time, it was touch and go, but Max—and blogging— have done well in the years since.

Along similar lines, several colleagues responded to the COVID-19 epidemic by collecting data from available sites and doing modeling and graphing. Professional organizations started making graphs available not long after, but these folks couldn't just sit on their hands until then. They wanted to *do something,* and programming is perfect for that.

Most programs are smaller scale but still serve some real need. For example, once my wife and I set out to lease a car. The car salesman gave me a worksheet with all sorts of numbers, including interest rates, money down, monthly payments, and the like. It didn't add up. Literally. I couldn't figure out where the numbers were coming from. I wrote my own lease calculator. It didn't actually help me understand how they calculate leases, but it did intimidate the salesman enough that he kept lowering the price to try to get me to stop. Score!

I've talked to people who write programs to split up a check, calculate their taxes, or even combine calculations with machine learning to figure out a fair price for a new home. For example, I recently heard some people arguing about calculating taxes on a gift. They were saying that giving someone of limited means a new car, as Oprah famously did, means that person would have to pay taxes on the car. Sure, that's a lot less than paying for that car, but sometimes it's still too much. So if you want the gift to be a true gift, you should give them the extra money they need to pay the taxes on the original gift amount. Ah, but that extra money is a gift, too. They'll need to pay taxes on that money, won't they? So maybe you should also give them the money to pay for that.

What happens if we play this all the way out? Do we end up paying an infinite amount of tax? It turns out we can solve a problem like this using algebra if we just think about it the right way. But it's arguably more natural to write a quick program. After all, the algebraic approach requires you to manipulate a bunch of symbols representing tax rates and gift amounts in a way that creates a mathematical expression that may not be interpretable. The programming approach directly emulates your thought process about the problem itself, paying the tax, then going back to pay the tax and the tax, and so forth. The relationship to the original problem description is more direct. Using the programming concepts we've discussed, we can write:

```
define the function total gift and taxes on gift:
  set total gifted to gift
  do forever:
    set tax on gift to gift x 12%
    if tax on gift is less than 0.01: break out of loop
    add tax on gift to total gifted
    set gift to tax on gift
  return total gifted
```

The looping structure here might look familiar. It's nearly the same as what Copilot proposed in chapter 8 for solving the rainfall problem introduced in chapter 2. It computes the tax on *gift*, then considers the resulting tax to be the new gift amount to be given, and so on until the calculated tax is below one cent. Each of the gifts is added into a *total gifted* amount, which is returned at the end. If we ask it to compute the total gift amount, including the tax on the tax on the tax and so on for a $10,000 car, it returns $11,363.64. Mighty helpful. A nice thing about this little program is that we can easily change the original gift amount, or even the stopping condition or the tax rate, and we still get our answers instantly.

Another category of solutions to everyday problems leverages this fast computation to do a kind of *reverse* calculation. Sticking with the gift example, let's say you are willing to make a gift of $15,000, but you want to make sure all the taxes on the taxes on the taxes and so on are covered. How pricey a car can you afford to give? Whereas the "forward" calculation we just went through from car cost to total cost follows pretty naturally from the way we think about taxes, it's not clear how to apply this idea in the backward direction to figure out the car cost from the total cost.

I can hear my colleague Ron Parr at Duke University say, "See, that's why you should have done the algebra. Algebra is reversible." True, but do we really have to think that hard? Couldn't we guess different values for the initial gift amount, see what the resulting total would be, then tweak the initial gift amount up or down accordingly? I tried that, and it took me only seven guesses to figure out that an initial gift of $13,200 results in a total, all-recursive-taxes-included gift of $15,000.

But it was still an awful lot of thinking to make all those guesses. Can't we automate that part, too? Indeed. The machine learning recipe we talked about can take care of this backward calculation for us. The representational space is the possible values of the gift, which for purposes of this example we'll set to dollar values between $0 and $15,000. The loss function could calculate the difference between the output of the _total gift and taxes_ function and the target output of $15,000. And the optimizer is . . . not important. You'd probably want to use the calculus-based process mentioned in chapter 5, but my point is that these sorts of decisions should be made automatically and shouldn't matter to us when solving problems like this. So, you ought to be able to say something like:

```
find a value for gift (between 0 and 15000) to minimize loss:
    set loss to
        difference between total gift and taxes on gift AND $15,000
```

For what it's worth, I think taxes on gifts, at least in the US, are generally paid by the gift giver and only kick in if the gift is worth more than $15,000 . . . but don't take tax advice from a computer science professor (even one who did a tax prep commercial at one point). So all of this is moot. But the same process can be used in many, many other problems, including creating budgets with a target dollar value or dividing up the cost of a shared purchase. It may even be useful for calculating a car lease, but I still don't get how those work.

Another application of the reverse calculation idea comes up all the time: scheduling. I've heard of several brides-to-be and a few grooms-to-be using this approach to help create seating charts for their wedding receptions. The basic idea is that it's relatively easy to write a program to check whether an allocation of guests to tables is okay. For example, it's bad to put Cousin Karen with Cousin Terry because a political argument will likely break out:

```
define the function political check on table assignment:
    if table assignment of Karen
        is the same as table assignment of Terry:
        return bad
    else if table assignment of Tim
        is the same as table assignment of Richard:
        return bad
    else:
        return good
```

Same thing with Richard and Tim. Otherwise, things are as good as could be expected.

In addition, every table should have eight to ten guests assigned to it:

```
define the function size check on table assignment:
    for each table in set of tables:
        set at table to 0
        for each guest in set of guests:
            if table assignment of guest is table:
                add 1 to at table
        if at table < 8 OR at table > 10:
            return bad
    return good
```

The function checks each table, counting up the number of guests assigned there. If it sees the total is not between eight and ten, it reports a bad assignment. If the loop makes it all the way through all the tables without noticing anything bad, it reports the assignment is good as far as table sizes are concerned.

Finally, everyone should be at a table where they know at least two other people. We need a separate function that encodes who knows whom, which I'll assume we already have in what follows:

```
define the function known check on table assignment:
    for each guest1 in set of guests:
        set know at table to 0
        for each guest2 in set of guests:
            if table assignment of guest1 is the same as
                    table assignment of guest2
                AND guest1 knows guest2:
                add 1 to know at table
```

```
if know at table < 2:
    return bad
return good
```

Like the previous function, if it finds someone sitting at a table with too few people he or she knows, the table assignment is bad. If we get through the whole list without finding such a person, the table assignment is good as far as this check is concerned.

The functions _known check_, _size check_, and _political check_ all do a forward calculation to tell if a table assignment is good or bad. Now, we can run them backward to find a table assignment that is good:

```
find a value for table assignment (from set of guests to set of tables)
    to satisfy:
  known check on table assignment is good
  AND size check on table assignment is good
  AND political check on table assignment is good
```

If the computer can find something that satisfies all the checks, that's a seating arrangement worth considering!

Now, if writing the program and creating the data about who knows whom seems harder than placing sticky notes on a big diagram of the tables, just remember Scott from high school. He's the guy who, at the very last minute, announces he has a new girlfriend and will only come if she is invited. Now you have to redo the whole assignment. In the world where you are using sticky notes, that's doubling your work. In the world where you are using a program, it's a simple matter of adding her name and who she knows to the program. The computer redoes the reverse calculation that produces the table assignments faster than you can call your other high school friends to complain about Scott.

For scheduling problems, the reverse calculation is typically performed by a _constraint satisfaction solver_. Like machine learning algorithms, these solvers have gotten amazingly good in the last decade or so. In addition to seating charts, folks are using them to help plan events that let people meet others outside their normal social group, schedule games in sports leagues, and assign workers to shifts to ensure coverage. It can be hard to keep all the various constraints in mind when you do these things by hand, so delegating the search to a machine can be a huge help.

A closely related task is having the computer help you out by randomizing. Being random is actually quite hard. People fall into patterns that cause them to make the same choices over and over again. Professional athletes, especially pitchers and tennis players who are trying to make their throws and shots hard to predict, get better and better at it over time, but most of us don't have the opportunity to perfect the skill of randomization and would benefit from automating the choices.

I've used the computer to simulate rolling dice or flicking a spinner for board games when we can't find the original equipment. But, a far more consequential case comes up when you are responsible for ensuring fairness in a group. For example, assistant professor Christian Muise told me he doesn't trust himself to call on his students in lab meetings in a way that's fair. It's just too easy to get into a rut and call on the same handful of students over and over again. So he wrote a simple script that generates a random order each time.

```
take a web page that lists all the lab members
pull out the name section of the web page
compile the name section into a list
for each name in the list:
    assign name a random number in a list of assigned numbers
sort the list by the assigned numbers
return the list
```

I often use the same idea when I'm reviewing papers or grading students to avoid always assessing the same people first, either because they hand in things early or they have a name like Aardvarkstein that comes at the beginning of the alphabet.

Other people write programs to support groups by combining randomization with the reverse calculation idea—assigning chores to share the load, or suggesting diverse restaurants to keep from always eating at the same two places, for example. Having the computer do the randomization also means you can keep things private if needed. Twitter respondent Bronson Harry told me he wrote a program to organize Secret Santa matchups to make sure they are random, don't repeat from year to year, and can be automatically shared by email so that everyone can participate in the surprise.

Speaking of heartwarming surprises, consider the tale Roy Keyes pointed me to on Twitter. His son fell in love with a puppy at a local shelter and desperately wanted to adopt it for his mom. The shelter said the puppy was still too young to adopt. They didn't have a waiting list, but proposed that Roy should just call each day to find out if the puppy was ready. But Roy feared the heartbreak of calling one day only to learn that someone else had picked up the puppy first. Instead, he wrote a program to check the shelter's web page and extract the current status, notifying him if it changed from "not available." He discovered that the tag "lblstage" was used in the formatting code of the web page to tag the status, so he wrote:

```
define the function schnoodle sniping:
    get the dog's webpage from its URL
    set status to the text marked with "lblstage" on the webpage
    if status is not "not available":
        send phone notification "Puppy may be available!"
```

He then set up trigger-action program (chapter 4) to run the function every fifteen minutes:

```
· every 15 minutes: run schnoodle sniping
```

A few days and a couple hundred check-ins later, Roy's phone pinged and the web page said the dog was available. I am happy to report that dog and family have been very happy together ever since.

I've heard similar stories about people writing code to bid in online auctions, to buy Taylor Swift tickets, to grab a slot for a COVID-19 vaccine, and to schedule a visa appointment to travel overseas. Basically, computers can wait in line for you when you're too busy or even when there's simply no line to wait in.

These examples are just a taste of what people will do in a future where everyone programs. Today, tapping into the 2 × 2 grid from chapter 1 to tell machines what to do by telling, explaining, demonstrating, and inspiring is still pretty tough. The biggest unknown is what happens when websites and services and even the objects we use every day can respond to our programming. Consider that jotting down a quick to-do list is super useful and, of course, requires that the to-do list writer can write. But it also requires that paper and pencils be ubiquitous: People are not that likely to fetch a chisel and waste a stone tablet just to remind themselves to shower and pick their kids up from school.

Universal programming begets universal programmability, and that's something we've never experienced.

CODE DEPENDENCY

Codependent relationships come in many forms. A stereotypical example is of a husband who is an alcoholic and a wife who ends up devoting her life to caring for him, ultimately blocking his opportunities to heal and her own opportunities for self-actualization: Their reliance on each other promotes unhealthy development for both.

Our relationship to computers has become codependent. What started as a process of software providers making computers easier to use for our own purposes has soured into us being spoon-fed information, unable to take initiative and instead spending detrimental amounts of time plugged in habitually checking for new updates.

Twelve-step programs for codependents sometimes make use of the "serenity prayer," which can be traced back to the writings of the eleventh-century Jewish philosopher Solomon Ibn Gabirol:

And they said: At the head of all understanding is realizing what is and what cannot be, and the consoling of what is not in our power to change.

In other words, there are things we can control and things we cannot control. Knowing which is which is essential.

In our current relationship with computers, knowing what we can and can't control has become way too hard, in part because computer apps generally don't give us ways to program them, and in small part because most of us don't know *how* to tell a computer what to do. Codependency for the loss!

Take magazines. In the old days, people were always trying to sell you subscriptions to magazines. On newsstands in the '80s, you could find *PC Magazine* targeted to computer users, *Teen*, *Mademoiselle*, *Self*, and *McCall's* to women of various ages, *Inside Sports* to fans, *Omni* to budding pseudoscientists, and *NASCAR Illustrated* to race car enthusiasts. As it turns out, all of these particular publications stopped being available in print since the rise of websites. But back then, if you were reading a story about the NFL quarterback Joe Montana (no relation to Hannah), it was probably because you picked up the relevant magazine. Picking what magazine to read doesn't seem like a profound act of self-determination. But it is a huge step beyond what we get when a computer takes over the choice. Now when you find a story in your newsfeed about a football player, you don't know if it's because the AI that organizes the feed thinks you're a sports fan, because it's national news that everyone is talking about, because it's sponsored by the league to get you to spend more on football, or because there's something political going on and it's being pushed as a wedge issue to change how you vote. It just seems like it's part of the zeitgeist and you accept it as such. When something appears out of the blue, not through our own conscious choice, it feels more inevitable, more universal, more broadly significant. Ultimately, that feeling of "not choosing" is warping our perception of the world and making it harder for us to exercise our own will.

At the end of the day, my biggest concern is that there's a fine line between making computers easier to use and making them *too* easy to use at the cost of user control. Programming of any sort is harder than not programming. It takes concentration and self-reflection. Most important, it requires that we figure out what we *want* the computer to do for us. That's not simple, but it is absolutely critical to keep us from becoming an extension of the machine—a resource that is kept in a passive state to be used by whomever takes the initiative to decide what we should be doing.

We're in a bit of a spot now. Asserting our will to the machine isn't just hard, it's harder than it used to be. Computing systems have become much more complex, and hooks into the code that could give us control are being hidden in the name of "simplicity" or "safety." The design of these systems isn't going to change until we make a concerted effort to push back. We need to insist on having more control before tech companies will even consider doing the work to give us any.

Here are some steps we can take:

1. Be reflective: When you are using a computer, ask yourself what you are trying to accomplish and try to stick with it. When you are not

using a computer, ask yourself whether a computer might be able to help automate your effort. Computers should empower our goals.

2. Insist on support: If there are things you want your computer to do but existing software won't allow it, push back. Ask the companies making the software to include the access you need. If other software is more amenable, switch. Companies should empower our goals.

3. Learn to program: If the software you are using supports it, make the effort to tailor the system to your needs. Ideally, future systems will meet us partway and this process will become easier, but only if the demand is there. Develop your ability to translate your ideas into a form that machines can carry out. Our skills should empower our goals.

There are some positive signs. Customizable questionnaires, web pages, and video games are out there now that allow people to write small amounts of code and get large benefits in terms of tailoring the computer's behavior to their goals. Word processors and spreadsheets provide scripting languages and macros that let you streamline your workflow. Entities like Google and Amazon are privately talking about giving people more control over how recommendations are made on their behalf, making it less about the machine reading our passive intentions and more about you asserting your will.

We need to do the work of asking ourselves what we want the computer to do for us, to learn skills for turning those intentions into actions, and to reflect on whether the resulting interaction is continuing to function as desired. It takes effort, but establishing a healthier relationship with our machines will make us more productive, more empowered, and, frankly, more joyful.

ACKNOWLEDGMENTS
THANKS!

I'd like to thank my wife, Lisa Littman, and kids, Max Littman and Molly Littman, for support and encouragement throughout this project. My parents, Phyllis Littman and Howard Littman, began providing their support and encouragement even earlier. All of them deserve credit for making me feel that teaching things, especially to one's kids, is valuable enough that we should remake our relationship with computers along the lines of parent–child relationships.

Lisa listened to several draft chapters read aloud. Molly went the extra mile and provided humor consultation for the entire manuscript. Gita Devi Manaktala and the MIT Press helped pinpoint what they thought were the topics to focus on. David Weinberger was amazingly thorough and educational in his role as lets-see-how-to-make-this-better editor.

I talked with a lot of people about the book while I was working on it, and those conversations shaped how I was thinking in ways both big and small. I want to express my appreciation to:

Dave Abel	Fiery Cushman
David Ackley	Sidney Davis
Stephen Bach	Jennifer Davis-Mitala
David Badre	Tina Eliassi-Rad
Justin Boyan	Jessica Forde
Ron Brachman	Jonathan Frankle
Michael Carbin	Eugene Freuder
Louis Castricato	Haotian Fu
Brian Christian	Thea Goldman
Jess Clark	Katherine Gorman

Chace Hayhurst
Mark Ho
Teri Howes
Charles Isbell
Sameen Jalal
Greg Keim
Elana Kimbrell
Jon Kleinberg
George Konidaris
Kweku Kwegyir-Aggrey
Lucas Lehnert
Jennifer Levy
Jill Littman
Sharon Lo
Eric Lupfer
Bertram Malle
Anita Nikolich
Neev Parikh
Hyojae Park
Gemima Philippe
Mark Riedl

Daniel Ritchie
Carolyn Rose
Lauren Schoenfeld
Steve Sloman
Bill Smart
Henry Sowerby
Hallie Stebbins
Aephraim Steinberg
Steven Strogatz
Kaushik Subramanian
Elizabeth Swayze
Bill Thomas
Blase Ur
Adrienne Urban
Andries van Dam
Jennifer Wang
Karen Wolin
Kyle Wray
Shangqun Yu
Lefan Zhang
Zhiyuan Zhou

The following Twitter folks kindly replied to my request for examples of how they use programming to solve problems in their day-to-day lives:

Austin Bowen (@AustinRBowen)
Bronson Harry (@BronsonBHarry)
Christian P. Hagen
   (@christianphagen)
Christian Muise (@cjmuise)
Stefano Ghirlanda (@drghirlanda)
Divya Shanmugam
   (@dmshanmugam)
10x engineer (@eng10x)
Frank Dellaert (@fdellaert)
Sam Aardvark (@HappyAar)

João G. M. Araújo (@_joaogui1)
Mitchel Kappen (@KappenMitchel)
Milad Khademi Nori
   (@khademinori)
Nathan Cahill (@NathanCahill20)
Rachel Lee (@rachel_s_l)
Ajey (@Paimaamu)
Patrick Mineault
   (@patrickmineault)
Roozbeh Mottaghi
   (@RoozbehMottaghi)

Roy Keyes (@roycoding)   Veronica (@veroniex_MM)

Saket Tiwari (@SaketTiwari14)   Chad C. Williams

Nick Sidorenko   (@WilliamsNeuro)

   (@NickSidorenkoCH)   Xipeng Wang (@xipengw1990)

* * *

INTERIOR. KITCHEN AREA OF ATLANTA APT.—EVENING

(MICHAEL is fussing with the dishwasher, pushing buttons on the control panel and wrinkling his nose. MOLLY is looking on, bleary-eyed.)

MOLLY: I don't know, Dad. You've tried everything. I think it's time for you to give up, at least for tonight. Just switch it off and we can order in Thai tomorrow.

MICHAEL: There's GOT to be a way to make the dishwasher stop beeping. Inside it's just a computer. Telling computers what I want them to do . . . that's kind of my superpower!

DISHWASHER: Beep.

MOLLY: Well, maybe you're telling it what you want but it can't DO it. *(MOLLY yawns.)* Like how I can't keep my eyes open. Listen. I'm going to bed. Peace out, Girl Scout.

MICHAEL: Fair enough, g'night, Molly Moo. But I'm just going to try one last thing before I pack it in. . . .

(MOLLY opens her mouth to argue, but decides against it. She exits, shaking her head skeptically. Her dog SADIE trots off after her.)

MICHAEL: If Molly's right, then maybe it's a hardware problem. Maybe I can route around it using this AI-dweeno board I picked up.

(MICHAEL wires the board to the dishwasher.)

DISHWASHER: Beep.

(MICHAEL makes a few adjustments.)

MICHAEL: There we go.

DISHWASHER: Beep.

MICHAEL: *(under his breath)* So, how can I get you to stop beeping?

DISHWASHER: *(in a British accent)* Permanently? Or only in particular circumstances?

(MICHAEL is shocked into silence.)

(Screen goes black. Then words appear: "THE DISHWASHER WILL RETURN.")

INDEX